UNDERSTANDING ○————
○THE NEW SOLAR
SYSTEM

UNDERSTANDING THE NEW SOLAR SYSTEM

FROM THE EDITORS OF *SCIENTIFIC AMERICAN*

Compiled and with introductions by

Sandy Fritz

Foreword by David H. Levy

A Byron Preiss Book

WARNER BOOKS

An AOL Time Warner Company

Copyright © 2002 by Scientific American, Inc., and
Byron Preiss Visual Publications, Inc.
All rights reserved.

The following essays first appeared in the pages of *Scientific American* magazine, as follows: "SOHO Reveals the Secrets of the Sun" by Kenneth R. Lang, March 1997. "The Pioneer Mission to Venus" by Janet G. Luhmann, James B. Pollack and Lawrence Colin, April 1994. "The Mars Pathfinder Mission" by Matthew P. Golombek, *Scientific American Presents*, 1999. "Migrating Planets" by Renu Malhotra, September 1999. "The Galileo Mission to Jupiter and Its Moons" by Torrence V. Johnson, February 2000. "Bejeweled Worlds" by Joseph A. Burns, Douglas P. Hamilton and Mark R. Showalter, February 2002. "Journey to the Farthest Planet" by S. Alan Stern, May 2002. "The Kuiper Belt" by Jane X. Luu and David C. Jewitt, May 1996. "The Oort Cloud" by Paul R. Weissman, September 1998. "Quest for the Limits of the Heliospere" by J.R. Jokipii and Frank B. McDonald, April 1995.

Warner Books, Inc., 1271 Avenue of the Americas, New York, NY 10020

Visit our Web site at www.twbookmark.com.

Printed in the United States of America
First Printing: December 2002
10 9 8 7 6 5 4 3 2 1

ISBN: 0-446-67953-4
Library of Congress Control Number: 2002108954

Ⓦ An AOL Time Warner Company

Cover design by j. vita
Book design by Gilda Hannah

Contents

Foreword

David H. Levy

ast night I set up a chair in the yard and looked up at the western sky. Five bright planets stared back, forming a tight grouping. Mercury was low in the west, flanked by a bright triangle of Venus, Mars, and Saturn. Much higher in the sky, Jupiter capped them all. These are the same five worlds that ancient cultures knew as wandering stars. Throughout most of recorded history, these wanderers were visualized to be circling the Earth. That changed in 1610, when Galileo pointed a telescope to Jupiter and discovered its four moons—smaller worlds that circle Jupiter and not the earth. A few months later, when Venus appeared in the evening sky and he began observing it regularly, Galileo discovered its changing phases. He could not interpret these phases in any way if Venus were circling the earth. However, he had no difficulty interpreting them if he assumed that Venus was circling the sun.

Thanks to his telescope, Galileo was the first to show the planets as orbiting the sun. He paid dearly for his conclusion. Summoned and tried before the Holy Office, Galileo was forced to recant, and to assert that the earth occupied the center of the solar system. If Galileo could come back for a visit somehow, imagine the surprise he would feel knowing that a mighty spacecraft bearing his name was now cruising through the Jupiter system, taking close-up images of the worlds he discovered so long ago. The spacecraft has observed volcanoes on Io, and a comet collision on Jupiter, items that stagger the mind and kindle the imagination.

Now would be a good time for Galileo to return, for we are undergoing a revolution in our understanding of the solar system that rivals what happened in Galileo's time. This second revolution actually began on October 4, 1959, when *Luna 3* began its journey from the Soviet Union to the moon. A few days later it photographed the far side, giving humanity its first view of a part of the Moon never seen before. The revolution entered a new phase three years later, when in December 1962 the American spacecraft *Mariner 2* reached Venus. Its successful journey was our first-ever visit to another planet. Since then we have seen triumph after triumph: the first human visit to the moon in July 1969; a robotic landing on Mars in 1976; the odyssey of *Voyager 2* to Jupiter, Saturn, Uranus, and Neptune; the *Magellan* probe of Venus; *Galileo's* tour of the Jupiter system—a list of events that hopefully will stretch out, like the line of Banquo's kings in Macbeth, to the crack of doom.

This revolution was helped along by Nature herself. On March 23, 1994, Carolyn and Gene Shoemaker and I took three pictures of a region of the sky near Jupiter, and discovered a fractured comet on those films. Comet Shoemaker-Levy 9, as it was called, had a few months earlier made an extremely close approach to Jupiter, close enough that Jupiter's tidal forces tore the comet into pieces. What we did not know that March was that the comet was headed directly to a collision with Jupiter in July 1994. During one incredible week, 21 fragments of the completely disrupted comet pummeled into Jupiter, yielding the largest planetary explosions ever seen by humanity.

The collisions were a watershed in our understanding of the solar system. In its earliest years, cometlike bodies called planetesimals hit each other to create and build up our system of planets. Comets persisted in colliding with the newly formed planets, bringing with them their precious organic materials and eventually setting off the origin of life, at least on Earth. Much later, a collision ended the era of the dinosaurs, opening the way for mammals and humanity. The breakup and collision of tiny comet Shoemaker-Levy 9 showed us how this process works.

What an amazing time to be alive! Our solar system is being studied in greater detail and in more ways than ever, and it has also provided its own road show on a scale never before seen. The book you have before you offers a way to understand and enjoy what is happening. I hope it will help you to find your way into humanity's new treasure trove of information and understanding of the true center of our universe—not our planet, but our entire solar family.

Introduction

Sandy Fritz

I t is from the perspective of our safe blue planet that we view with awe and wonder the night skies. Hundreds of thousands of points of lights twinkle overhead. Some are faraway stars, some form the familiar constellations, and some are reflections of the solar light from much closer celestial bodies that also orbit our sun. They are the companion planets of Earth in the solar system.

Each of the nine planets has unique qualities that are revealed in how each one responds to the input of solar energy. Mercury maintains a powerful magentosphere that partly protects its surface from the raw power of the sun. Venus, with its unique clockwise spin, heats up to 450 degrees Celsius at the surface and cycles sulfuric acid through its atmosphere. Earth seems to enjoy an ideal environment for life as we know it. The four inner planets are solid and rocky in construction, with molten cores. But it is the outer planets that draw the attention of most scientists who study the solar system today.

The outer planets have an entirely different character from the inner planets. Saturn, Jupiter, Uranus and Neptune are sometimes referred to as "the gas giants." These planets incorporate much more of the gaseous elements in their structures, as opposed to the inner, rocky planets. Jupiter and its network of moons can almost be considered a solar system unto itself. It produces more energy than it receives from the sun, and it has captured an impressive array of orbiting bodies. The larger of Jupiter's moons rival Mercury in size.

The four gas giants also sport ring formations. A delicate interplay between a planet's gravity, its orbit, and its distance from the sun gives rise to these structures. No two planetary ring systems are exactly alike, and none are static.

The farthermost known planet in the solar system may actually be a binary system. Pluto and its moon Charon are nearly the same size, and the pair's erratic orbit around the sun sometimes takes it inside the more orderly orbit of Neptune, and sometimes outside of it. Pluto has recently been suspected of belonging to a belt of smaller pieces of rock and "dirty snowballs" that were swept to the outer reaches of the solar system over time. The Kuiper belt, the innermost fringe of which is patrolled by Neptune, may be the origin of some of the comets that sweep through the solar system. Perhaps other Pluto-sized bodies are harbored within it, out of reach of our telescopes. An even farther ring of matter may gather around the distant fringes of the solar system. The Oort cloud, still undetected but predicted by theory, may also harbor comets and rocky bodies.

Through all these planets, satellites, belts, and clouds, the sun's radiation and its solar wind keep at bay the pressure of interstellar space. Yet somewhere there must be a boundary between our solar system and the rest of space. No one knows where these boundaries lie, but four aging spacecraft—a pair of *Pioneers* and a pair of *Voyagers*—are coasting to these frontiers, sending back trickles of data as they go. And with our ground-based radar dishes we follow them, curious about where our solar system begins, and where it ends.

The nearest star to Earth bathes our planet with the raw energy needed to make terrestrial life possible. The Solar Heliospheric Observatory (SOHO), launched in December 1995, has provided an unparalleled glimpse into the interior and the exterior of Sol, detailing the dynamics that give rise to a star's life.

SOHO Reveals the Secrets of the Sun

Kenneth R. Lang

F rom afar, the sun does not look very complex. To the casual observer, it is just a smooth, uniform ball of gas. Close inspection, however, shows that the star is in constant turmoil—a fact that fuels many fundamental mysteries. For instance, scientists do not understand how the sun generates its magnetic fields, which are responsible for most solar activity, including unpredictable explosions that cause magnetic storms and power blackouts here on the earth. Nor do they know why this magnetism is concentrated into so-called sunspots, dark islands on the sun's surface that are as large as the earth and thousands of times more magnetic. Furthermore, physicists cannot explain why the sun's magnetic activity varies dramatically, waning and intensifying again every 11 years or so.

To solve such puzzles—and better predict the sun's impact on our planet—the European Space Agency and the National Aeronautics and Space Administration (NASA) launched the two-ton Solar and Heliospheric Observatory (SOHO, for short) on December 2, 1995. The spacecraft reached its permanent strategic position—which is called the inner Lagrangian point and is about 1 percent of the way to the sun—on February 14, 1996. There SOHO is balanced between the pull of the earth's gravity and the sun's gravity and so orbits the sun together with the earth. Earlier spacecraft studying the sun orbited the earth, which would regularly obstruct their view. In contrast, SOHO monitors the sun continuously: 12 instruments examine the sun in unprecedented detail. They downlink several thousand images a day through NASA's Deep Space Network antennae to SOHO's Experimenters' Operations Facility at the NASA Goddard Space Flight Center in Greenbelt, Md.

At the Experimenters' Operations Facility, solar physicists from around the world work together, watching the sun night and day from a room without windows. Many of the unique images they receive move nearly instantaneously to the SOHO home page on the World Wide Web (http://sohowww.nascom. nasa.gov). When these pictures first began to arrive, the sun was at the very bottom of its 11-year activity cycle. But SOHO carries enough fuel to continue operating for a decade or more. Thus, it will keep watch over the sun through all its tempestuous seasons. Already, though, SOHO has offered some astounding findings.

Exploring Unseen Depths

To understand the sun's cycles, we must look deep inside the star, to where its magnetism is generated. One way to explore these unseen depths is by tracing the in-and-out, heaving motions of the sun's outermost visible surface, named the photosphere from the Greek word *photos*, meaning "light." These oscillations, which can be tens of kilometers high and travel a few hundred meters per second, arise from sounds that course through the solar interior. The sounds are trapped inside the sun; they cannot propagate through the near vacuum of space. (Even if they could reach the earth, they are too low for human hearing.) Nevertheless, when these sounds strike the sun's surface and rebound back down, they disturb the gases there, causing them to rise and fall, slowly and rhythmically, within a period of about five minutes.

The throbbing motions these sounds create are imperceptible to the naked eye, but SOHO instruments routinely pick them out. Two devices, the Michelson Doppler Imager (MDI) and the Global Oscillations at Low Frequencies (GOLF), detect surface oscillation speeds with remarkable precision—to better than one millimeter per second. A third device tracks another change the sound waves cause: as these vibrations interfere with gases in light-emitting regions of the sun, the entire orb flickers like a giant strobe. SOHO's Variability of solar IRradiance and Gravity Oscillations (VIRGO) device records these intensity changes, which are but minute fractions of the sun's average brightness.

The surface oscillations are the combined effect of about 10 million separate notes—each of which has a unique path of propagation and samples a well-defined section inside the sun. So to trace the star's physical landscape all the way through—from its churning convection zone, the outer 28.7 percent (by radius), into its radiative zone and core—we must determine the precise pitch of all the notes.

The dominant factor affecting each sound is its speed, which in turn depends on the temperature and composition of the solar regions through which it passes. SOHO scientists compute the expected sound speed using a numerical model. They then use relatively small discrepancies between their computer calculations and the observed sound speed to fine-tune the model and establish the sun's radial variation in temperature, density and composition.

At present, theoretical expectations and observations made

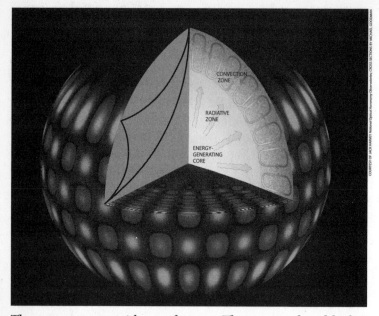

The sun resonates with sound waves. They are produced by hot gas churning in the convection zone sandwiched between the sun's core and its surface. Acoustic waves traveling out toward the surface are reflected back toward the core. The core, in turn, reflects the waves back toward the surface.

with the MDI telescope are in close agreement, showing a maximum difference of only 0.2 percent. Where these discrepancies occur is, in fact, significant. They suggest that material is mixing at the boundary of the energy-generating core and also just below the convection zone.

For more than three centuries, astronomers have known from watching sunspots that the photosphere rotates faster at the equator than at higher latitudes and that the speed decreases evenly toward each pole. SOHO data confirm that this differential pattern persists through the convection zone. Furthermore, the rotation speed becomes uniform from pole to pole about a third of the way down. Thus, the rotation velocity changes sharply at the base of the convection zone. There the outer parts of the radiative interior, which rotates at one speed, meet the overlying convection zone, which spins faster in its equatorial middle. We now suspect that this thin base layer of rotational shear may be the source of the sun's magnetism.

The MDI telescope on board SOHO has also helped probe the sun's outer shells. Because its lenses are positioned well above the earth's obscuring atmosphere, it can continuously resolve fine detail that cannot always be seen from the ground. For this reason, it has proved particularly useful in time-distance helioseismology, a new technique for revealing the motion of gases just below the photosphere. The method is quite straightforward: the telescope records small periodic changes in the wavelength of light emitted from a million points across the sun every minute. By keeping track of them, it is possible to determine how long it takes for sound waves to skim through the sun's outer layers. This travel time tells of both the temperature and gas flows along the internal path connecting two points on the visible solar surface. If the local temperature is high, sound waves move more quickly—as they do if they travel with the flow of gas.

The MDI has provided travel times for sounds crossing thousands of paths, linking myriad surface points. And SOHO scientists have used these data to chart the three-dimensional internal structure and dynamics of the sun, much in the same way that a computed tomographic (CT) scan creates an image of the inside of the brain. They fed the SOHO data to supercomputers to work out temperatures and flow directions along these intersecting paths. After a solid week of number crunching, the machines generated the first maps showing convective flow velocities inside a star. These flows are not global motions, such as rotations, but rather small-scale ones that seem to be independent of one another. Even so, their speed reaches one kilometer per second—which is faster than a supersonic jet airplane.

To get a look at these flows diving down through the convection zone, the MDI team computed travel times for sounds moving some 8,000 kilometers down into the sun. The researchers found that, as expected, this tumultuous region resembles a pot of boiling water: hot material rises through it, and cooler gases sink. Many of these flows are, however, unexpectedly shallow. The team also investigated horizontal motions at a depth of about 1,400 kilometers and compared them with an overlying magnetic image, also taken by the MDI instrument. They found that strong magnetic concentrations tend to lie in regions where the subsurface gas flow converges. Thus, the churning gas probably forces magnetic fields together and concentrates them, thereby overcoming the outward magnetic pressure that ought to make such localized concentrations expand and disperse.

SOHO is also helping scientists explain the solar atmosphere, or corona. The sun's sharp outer rim is illusory. It merely marks the level beyond which solar gas becomes transparent. The invisible corona extends beyond the planets and presents one of the most puzzling paradoxes of solar physics: it is unexpectedly hot, reaching temperatures of more than one million kelvins just above the photosphere; the sun's visible surface is only 5,780 kelvins. Heat simply should not flow outward from a cooler to a hotter region. It violates the second law of thermodynamics and all common sense as well. Thus, there must be some mechanism transporting energy from the photosphere, or below, out to the corona. Both kinetic and magnetic energy can flow from cold to hot regions. So writhing gases and shifting magnetic fields may be accountable.

For studying the corona and identifying its elusive heating mechanism, physicists look at ultraviolet (UV), extreme ultraviolet (EUV) and x-ray radiation. This is because hot material—such as that within the corona—emits most of its energy at these wavelengths. Also, the photosphere is too cool to emit intense radiation at these wavelengths, so it appears dark under the hot gas. Unfortunately, UV, EUV and x-rays are partially or totally absorbed by the earth's atmosphere, and so they

SOHO's Instruments

Researchers around the world are studying the sun using 12 instruments on board SOHO. Three devices probe the sun's interior; six measure the solar atmosphere; and three keep track of the star's far-reaching winds.

INSTRUMENT	Measurement	PRINCIPAL INVESTIGATOR
GOLF	The Global Oscillations at Low Frequencies device records the velocity of global oscillations within the sun	Alan H. Gabriel, Institut d'Astrophysique Spatiale, France
VIRGO	The Variability of solar IRradiance and Gravity Oscillations instrument measures fluctuations in the sun's brightness, as well as its precise energy output	Claus Fröhlich, Physico Meteorological Observatory Davos and World Radiation Center, Switzerland
SOI/MDI	The Solar Oscillations Investigation/ Michelson Doppler Imager measures the velocity of oscillations, produced by sounds trapped inside the sun	Phillip H. Scherrer, Stanford University, U.S.
SUMER	The Solar Ultraviolet Measurements of Emitted Radiation instrument gives data about the temperatures, densities and velocities of various gases in the chromosphere and corona	Klaus Wilhelm, Max Planck Institute for Aeronomy, Germany
CDS	The Coronal Diagnostic Spectrometer records the temperature and density of gases in the corona	Richard A. Harrison, Rutherford Appleton Laboratory, U.K.
EIT	The Extreme-ultraviolet Imaging Telescope provides full-disk images of the chromosphere and the corona	Jean-Pierre Delaboudinière Institut d'Astrophysique Spatiale
UVCS	The UltraViolet Coronagraph Spectrometer measures the temperatures and velocities of hydrogen atoms, oxygen and other ions in the corona	John L. Kohl, Smithsonian Astrophysical Observatory, U.S.
LASCO	The Large Angle Spectroscopic COronograph provides images that reveal the corona's activity, mass momentum and energy	Guenter E. Brueckner, Naval Research Laboratory, U.S.
SWAN	The Solar Wind ANisotropies device monitors latititudinal and temporal variations in the solar wind	Jean-Loup Bertaux, Service d'Aéronomie France
CELIAS	The Charge, ELement and Isotope Analysis System quantifies the mass, charge, composition and energy distribution of particles in the solar wind	Peter Bochsler, University of Bern, Switzerland
COSTEP	The COmprehensive SupraThermal and Energetic Particle analyzer determines the energy distribution of protons, helium ions and electrons	Horst Kunow, University of Kiel, Germany
ERNE	The Energetic and Relativistic Nuclei and Electron experiment measures the energy	Jarmo Torsti, University of Turku,

must be observed through telescopes in space. SOHO is now measuring radiation at UV and EUV wavelengths using four instruments: the Extreme-ultraviolet Imaging Telescope (EIT), the Solar Ultraviolet Measurements of Emitted Radiation (SUMER), the Coronal Diagnostic Spectrometer (CDS) and the UltraViolet Coronagraph Spectrometer (UVCS).

To map out structures across the solar disk, ranging in temperature from 6,000 to two million kelvins, SOHO makes use of spectral lines. These lines appear when the sun's radiation intensity is displayed as a function of wavelength. The various SOHO instruments locate regions having a specific temperature by tuning into spectral lines emitted by the ions formed there. Atoms in a hotter gas lose more electrons through collisions, and so they become more highly ionized. Because these different ions emit spectral lines at different wavelengths, they serve as a kind of thermometer. We can also infer the speed of the material moving in these regions from the Doppler wavelength changes of the spectral lines SOHO records.

Ultraviolet radiation has recently revealed that the sun is a vigorous, violent place even when its 11-year activity cycle is in an apparent slump—and this fact may help explain why the corona is so hot. The whole sun seems to sparkle in the UV light emitted by localized bright spots. According to SOHO measurements, these ubiquitous hot spots are formed at a temperature of a million kelvins, and they seem to originate in small, magnetic loops of hot gas found all over the sun, including both its north and south poles. Some of these spots explode and hurl material outward at speeds of hundreds of kilometers per second. SOHO scientists are now studying these bright spots to see if they play an important role in the elusive coronal heating mechanism.

To explore changes at higher levels in the sun's atmosphere, SOHO relies on its UVCS and its Large Angle Spectroscopic COronagraph (LASCO). Both instruments use occulting disks to block the photosphere's underlying glare. LASCO detects visible sunlight scattered by electrons in the corona. Initially it revealed a simple corona—one that was highly symmetrical and stable. This corona, viewed during the sun's magnetic lull, exhibited pronounced holes in the north and south. (Coronal holes are extended, low-density, low-temperature regions where EUV and x-ray emissions are abnormally low or absent.)

In contrast, the equatorial regions were ringed by straight, flat streamers of outflowing matter. The sun's magnetic field shapes these streamers. At their base, electrified matter is densely concentrated within magnetized loops rooted in the photosphere. Farther out in the corona, the streamers narrow and stretch tens of millions of kilometers into space. These extensions confine material at temperatures of about two million kelvins within their elongated magnetic boundaries, creating a belt of hot gas that extends around the sun.

The streamers live up to their name: material seems to flow continuously along their open magnetic fields. Occasionally the coronagraphs record dense concentrations of material moving through an otherwise unchanging streamer—like seeing leaves floating on a moving stream. And sometimes tremendous eruptions, called coronal mass ejections, punctuate the steady outward flow. These ejections hurl billions of tons of million-degree gases into interplanetary space at speeds of hundreds of kilometers per second. This material often reaches the earth in only two or three days. To almost everyone's astonishment, LASCO found equatorial ejections emitted within hours of each other from opposite sides of the sun.

The coronagraphs have only a side view of the sun and so can barely see material moving to or from the earth. But based on what we can see, we guess that these ejections are global disturbances, extending all the way around the sun. In fact, unexpectedly wide regions of the sun seem to convulse when the star releases coronal mass ejections, at least during the minimum in the 11-year activity cycle. And the coronagraph has detected that a few days before the ejections, the streamer belt gets brighter, suggesting that more material is accruing there. The pressure and tension of this added material probably build until the streamer belt blows open in the form of an ejection. The entire process is most likely related to a large-scale global reorganization of the sun's magnetic field.

The sun's hot and stormy atmosphere is forever expanding in all directions, filling the solar system with a ceaseless flow—called the solar wind—that contains electrons, ions and magnetic fields. The million-degree corona creates an outward pressure that overcomes the sun's gravitational attraction, enabling this perpetual outward flow. The wind accelerates as it moves away from the sun, like water overflowing a dam. As the corona disperses, it must be replaced by gases welling up from below to feed the wind. Earlier spacecraft measurements, as well as those from Ulysses (launched in 1990), showed that the wind has a fast and a slow component. The fast one moves at about 800 kilometers per second; the slow one travels at half that speed.

No one knows exactly where the slowspeed component originates or what gives the high-speed component its additional push, but SOHO should provide the answers. The slow component is associated with equatorial regions of the sun, now being scrutinized by LASCO and UVCS. The high-speed component pours forth from the polar coronal holes. (Open magnetic fields there allow charged particles to escape the sun's gravitational and magnetic grasp.) SOHO is now investigating whether polar plumes—tall structures rooted in the photosphere that extend into the coronal holes—help to generate this high-speed solar wind.

SOHO's UVCS has examined the spectral emission of hydrogen and heavily charged oxygen ions in the regions where the corona is heated and the solar wind accelerates. And these spectral-line profiles have produced surprising results, revealing a marked difference in the agitation speeds at which hydrogen and oxygen ions move. In polar coronal holes, where the fast solar wind originates, the heavier oxygen is far more agitated, with about 60 times more energy of motion; above two solar radii from the sun's center, oxygen has the higher agitation speed, approaching 500 kilometers per second. Hydrogen, on the other hand, moves at only 250 kilometers per second. In contrast, within equatorial regions, where the slow-speed wind begins, the lighter hydrogen moves faster than the oxygen, as one would expect from a heat-driven wind.

Researchers are now trying to determine why the more massive oxygen ions move at greater speeds in coronal holes. Information about the heating and acceleration processes is probably retained within the low-density coronal holes, wherein ions rarely collide with electrons. Frequent collisions in high-density streamers might erase any signature of the relevant processes.

Another instrument on board SOHO, the Solar Wind ANisotropies (SWAN), examines interstellar hydrogen atoms sweeping through our solar system from elsewhere. The sun's ultraviolet radiation illuminates this hydrogen, much the way that a street lamp lights a foggy mist at night. The solar wind particles tear the hydrogen atoms apart. For this reason, where the wind passes through the interstellar hydrogen cloud, it creates a dark cavity in its wake. The ultraviolet glow detected by this instrument therefore outlines the shape of the sun's wind. So far these measurements indicate that the solar wind is more intense in the equatorial plane of the sun than over the north or south poles.

As our civilization becomes increasingly dependent on sophisticated systems in space, it becomes more vulnerable to sun-driven space weather. In addition to magnetic storms and power surges, forceful coronal mass ejections can trigger intense auroras in the polar skies and damage or destroy earth-orbiting satellites. Other intense eruptions, known as solar flares, hurl out energetic particles that can endanger astronauts and destroy satellite electronics. If we knew the solar magnetic changes preceding these violent events, then SOHO could provide the early warning needed to protect us from their effects.

Indeed, parked just outside the earth, SOHO can sample threatening particles before they get to us. SOHO's Charge, ELement and Isotope Analysis System (CELIAS) currently measures the abundance of rare elements and isotopes that were previously not observable. By comparing these, we can reach certain conclusions about conditions, in the sun's atmosphere, where the solar wind originates. Two other instruments, the COmprehensive SupraThermal and Energetic Particle (COSTEP) analyzer and the Energetic and Relativistic Nuclei and Electron experiment (ERNE), have already obtained direct in situ measurements of very energetic electrons, protons and helium nuclei approaching the earth. They traced them back to violent eruptions detected by the EIT at the sun. Such events will surely become more numerous as we enter the next maximum in solar activity. And then SOHO will be able to follow such eruptions as they begin below the sun's visible surface and travel through the sun's atmosphere to affect the earth and the rest of the solar system.

SOHO has obtained marvelous results to date. It has revealed features on the mysterious sun never seen before or never seen so clearly. It has provided new insights into fundamental unsolved problems, all the way from the sun's interior to the earth and out to the farthest reaches of the solar wind. Some of its instruments are now poised to resolve several other mysteries. Two of them, the GOLF and VIRGO instruments, will soon have looked at the solar oscillations long enough, and deep enough, to determine the temperature and rotation at the sun's center. Moreover, during the next few years, our home star's inner turmoil and related magnetic activity—which can directly affect our daily lives—will increase. SOHO should then offer even greater scientific returns, determining how its threatening eruptions and hot, gusty winds originate and perhaps predicting conditions in the sun's atmosphere.

During its 14-year mission, the Pioneer Venus Orbiter *stripped many of the secrets from this cloud-enshrouded planet. Rather than being "Earth's Twin" as was thought in an earlier century, Venus is now recognized to be a planet where clouds of sulfur dioxide rain sulfuric acid on a surface hot enough to melt lead.*

The *Pioneer* Mission to Venus

Janet G. Luhmann, James B. Pollack and Lawrence Colin

Venus is sometimes referred to as the earth's "twin" because it resembles the earth in size and in distance from the sun. Over its 11 years of operation, NASA's *Pioneer Venus* mission revealed that the relation between the two worlds is more analogous to Dr. Jekyll and Mr. Hyde. The surface of Venus bakes under a dense carbon dioxide atmosphere, the overlying clouds consist of noxious sulfuric acid, and the planet's lack of a magnetic field exposes the upper atmosphere to the continuous hail of charged particles from the sun. Our opportunity to explore the hostile Venusian environment came to an abrupt close in October 1992, when the *Pioneer Venus Orbiter* burned up like a meteor in the thick Venusian atmosphere. The craft's demise marked the end of an era for the U.S. space program.

The information gleaned by *Pioneer Venus* complements the well-publicized radar images recently sent back by the *Magellan* spacecraft. *Magellan* concentrated on studies of Venus's surface geology and interior structure. *Pioneer Venus*, in comparison, gathered data on the composition and dynamics of the planet's atmosphere and interplanetary surroundings. These findings illustrate how seemingly small differences in physical conditions have sent Venus and the earth hurtling down very different evolutionary paths. Such knowledge will help scientists intelligently evaluate how human activity may be changing the environment on the earth.

Pioneer Venus consisted of two components, the *Orbiter* and the *Multiprobe*. The *Multiprobe* carried four craft (one large probe and three small identical ones) designed to plunge into the Venusian atmosphere, sending back data on the local conditions along the way. The *Orbiter* bristled with a dozen instruments with which to examine the composition and physical nature of Venus's upper atmosphere and ionosphere, the electrically charged layer between the atmosphere and outer space.

The *Multiprobe* was launched in August 1978 and reached Venus on December 9 of that year. Twenty-four days before its arrival, the *Multiprobe* carrier, or "bus," released the large probe; about five days later the bus freed the three small probes to begin their own, independent courses. The probes approached the planet from both high latitudes and low ones and from both the daylit and nighttime sides. In this way, information relayed by the probes during their descents enabled scientists to piece together a comprehensive picture of the atmospheric structure of Venus.

The *Orbiter* left the earth in May 1978, but it followed a longer trajectory than the *Multiprobe*, so it arrived only five days earlier, on December 4. At that time, the spacecraft entered a highly eccentric orbit that looped to within 150 to 200 kilometers from the planet's surface but carried it out to a distance of 66,900 kilometers. During its closest approaches to Venus, the *Orbiter*'s instruments could directly sample the planet's ionosphere and upper atmosphere. Twelve hours later the *Orbiter* would have receded far enough from Venus so that the craft's remote-sensing equipment could obtain global images of the planet and could measure its near-space environment.

The gravitational pull of the sun acted to change the shape of the probe's orbit. Starting in 1986, solar gravity caused the *Orbiter* to pass ever closer to the planet. When the spacecraft's thrusters ran out of fuel, *Pioneer Venus* dove deeper into the Venusian atmosphere on each orbit until it met its fiery end.

Well before the arrival of *Pioneer Venus*, astronomers had learned that Venus does not live up to its image as the earth's near twin. Whereas the earth maintains conditions ideal for liquid water and life, Venus is the planetary equivalent of hell. Its surface temperature of 450 degrees Celsius is hotter than the melting point of lead. Atmospheric pressure at the ground is some 93 times that at sea level on the earth.

Even aside from the heat and the pressure, the air on Venus would be utterly unbreathable to humans. The earth's atmosphere is about 78 percent nitrogen and 21 percent oxygen. Venus's much thicker atmosphere, in contrast, is composed almost entirely of carbon dioxide. Nitrogen, the next most abundant gas, makes up only about 3.5 percent of the gas molecules. Both planets possess about the same total amount of gaseous nitrogen, but Venus's atmosphere contains some 30,000 times as much carbon dioxide as does the earth's. In fact, the earth does hold a quantity of carbon dioxide comparable to that in the Venusian atmosphere. On the earth, however, the carbon dioxide is locked away in carbonate rocks, not in gaseous form in the air. This crucial distinction is responsible for many of the drastic environmental differences that exist between the two planets.

The large *Pioneer Venus* atmospheric probe carried a mass spectrometer and gas chromatograph, devices that measured the exact composition of the atmosphere of Venus. One of the most stunning aspects of the Venusian atmosphere is that it is extremely dry. It possesses only a hundred thousandth as much water as the earth has in its oceans. If all of Venus's water could somehow be condensed onto the surface, it would make

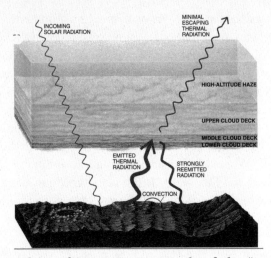

The atmosphere of Venus is an example of the "greenhouse effect" run wild. Incoming solar energy penetrates the planet's opaque clouds and strikes the surface. Thermal radiation from the planet's surface is reflected back to the ground by a dense lower cloud deck, resulting in surface temperatures of 450 degrees Celsius.

a global puddle only a couple of centimeters deep.

Unlike the earth, Venus harbors little if any molecular oxygen in its lower atmosphere. The abundant oxygen in the earth's atmosphere is a by-product of photosynthesis by plants; if not for the activity of living things, the earth's atmosphere also would be oxygen poor. The atmosphere of Venus is far richer than the earth's in sulfur-containing gases, primarily sulfur dioxide. On the earth, rain efficiently removes similar sulfur gases from the atmosphere.

Minor constituents of the Venusian atmosphere that were detected by *Pioneer Venus* offer clues about the internal history of the planet. The inert gas argon 40, for instance, is produced by the decay of radioactive potassium 40, which is present in nearly all rocks. As a planet's interior circulates, argon 40 that is trapped in deep rocks works its way to the surface and into the atmosphere, where it accumulates over the eons. *Pioneer Venus* found significantly less argon 40 in Venus's atmosphere than exists in the earth's. That disparity reflects a profound difference in how mass and heat are transported from each planet's interior to its surface. *Magellan* recently found evidence of earlier widespread volcanism on Venus but no signatures of the plate tectonics that keep the earth's surface geologically active and young.

Pioneer Venus revealed other ways in which Venus is more primeval than the earth. Venus's atmosphere contains higher concentrations of inert, or noble, gases—especially neon and other isotopes of argon—that have been present since the time the planets were born. This difference suggests that Venus has held on to a far greater fraction of its earliest atmosphere. Much of the earth's primitive atmosphere may have been stripped away and lost into space when our world was struck by a Mars-size body. Many planetary scientists now think the moon formed out of the cloud of debris that resulted from such a gigantic impact.

Venus's thick, carbon dioxide-dominated atmosphere is directly responsible for the inhospitable conditions on the planet's surface. On an airless body like the moon, the surface temperature depends simply on the balance between the amount of sunlight the ground absorbs and the amount of heat it emits back into space. The presence of an atmosphere complicates the situation. An atmosphere blocks some sunlight from reaching the surface and helps to carry heat upward. But more significantly, the atmospheric gases absorb infrared (thermal) radiation from the ground and reemit it back. The resultant warming of the surface is called the greenhouse effect because the atmosphere functions like a greenhouse: sunlight can get in, but infrared rays cannot get out, causing temperatures to rise.

The intensity of the greenhouse effect depends on how thoroughly the atmospheric gases capture infrared radiation.

The principal greenhouse gases on the earth—carbon dioxide and water vapor—absorb complementary parts of the infrared spectrum. Adding more of these gases to the air would, in theory, increase the efficiency of the greenhouse effect, which is why people worry about the climatic impact of carbon dioxide released by human activities. The earth's atmosphere is largely transparent to infrared rays having wavelengths between eight and 13 microns, or millionths of a meter (although ozone, methane, freon and other gases do absorb rays in narrow portions of this band). This open "window" in the atmospheric greenhouse limits the amount of warming that the earth experiences.

Pioneer Venus showed that the greenhouse effect operates much more efficiently on Venus. Data from the four atmospheric probes enabled workers to construct a mathematical model that closely matches the observed temperatures at various altitudes. From that model, it was deduced that carbon dioxide is the most significant greenhouse gas on Venus but that its action is enhanced by the presence of water vapor, clouds, sulfur dioxide and carbon monoxide. The mixture of gases and particles in the Venusian atmosphere blocks thermal radiation at virtually all wavelengths, preventing heat from escaping into space and yielding torrid surface temperatures. These results emphasize the importance of learning more about how human-generated greenhouse gases might affect the terrestrial climate.

Astronomers have long wondered how Venus turned out so hot and dry compared with the earth, especially given that Venus and the earth probably started out with similar overall compositions. According to present theory, the two planets grew by colliding with and absorbing smaller bodies. In the process, each protoplanet would have scattered some smaller bodies into orbits that would have crossed the other protoplanet's path. Hence, the earth and Venus should have accumulated comparable quantities of water-rich bodies even if, at first, water was irregularly distributed through the infant solar system. The roughly equal quantities of carbon dioxide and nitrogen on the two planets support the notion that they once had comparable amounts of water as well.

The young earth and Venus quickly developed thick atmospheres consisting of gases expelled from their interiors and of the vaporized remains of icy, impacting bodies. Water in the earth's atmosphere condensed into lakes and oceans, which proved crucial to the planet's climatic development. Much of the airborne carbon dioxide was quickly sequestered into solid carbonate, a process that occurs through the chemical weathering of rocks in the presence of liquid water.

Venus, too, may have had broad oceans during its youth. The newborn sun was about 30 percent less luminous than it is at present, so temperatures on Venus could have been well below the boiling point of water. (Venus orbits at 0.72 times the earth's distance from the sun.) As the sun brightened, however, surface temperatures on Venus eventually rose above boiling. From then on, any carbon dioxide exhaled by volcanoes or delivered by impacts on Venus could no longer be removed from the atmosphere by chemical weathering. As carbon dioxide accumulated in the atmosphere, the greenhouse effect grew ever more intense. The ultimate result was the sizzling, carbon dioxide dominated world of today.

After the oceans boiled, the atmosphere of Venus should have been full of water vapor—in clear contrast to the data. Where has all the water gone?

Pioneer Venus has helped answer that question. The spacecraft documented that even now Venus continues to lose water. Water molecules that wander above the cloud tops react with solar radiation and other molecules. In the process the water molecules split into their oxygen and hydrogen components. The lightweight hydrogen atoms may escape into space by interacting with energetic atoms and ions in the upper atmosphere or with the solar wind, a flow of charged particles that issues from the sun. The leftover oxygen atoms may combine with minerals on the surface, or they, too, may escape by interacting with the solar wind.

A few billion years ago Venus's upper atmosphere contained much more water than it does now; the early sun also emitted far more energetic ultraviolet rays. Both factors greatly hastened the rate at which Venus's water was destroyed and carried off into space. Calculations indicate that over the 4.5-billion-year lifetime of the solar system, Venus could have lost as much water as resides in the earth's oceans.

The earth never experienced such large losses of water because of its moderate surface temperature. Water on the earth stays mostly on the ground or in the lower atmosphere; very little reaches the upper atmosphere, where it may disappear forever. Once the oceans of Venus boiled, in contrast, the planet's atmosphere grew ever hotter, driving more and more water vapor into the upper reaches of the atmosphere.

And yet some water remains. Observations of Venus's upper atmosphere made by the *Pioneer Venus Orbiter* imply that the planet now loses about 5×10^{25} hydrogen atoms and ions each second. At that rate, the entire amount of water in the atmosphere would be gone in about 200 million years. Venus is more than 20 times that old, so some mechanism must replenish the water that Venus is constantly losing. The water most likely derives from a mix of external sources (such as the impact of comets and icy asteroids) and internal ones (through volcanic eruptions or more widespread and steady outgassing to the surface). Because the understanding of Venus's water loss is still quite sketchy, it is possible that *Pioneer Venus* might actually have observed the last trickle from the planet's water-rich early atmosphere.

Despite its lack of water, Venus is cloaked in thick clouds that conceal its surface from conventional telescopes. The nature of those clouds has intrigued astronomers for centuries. By the

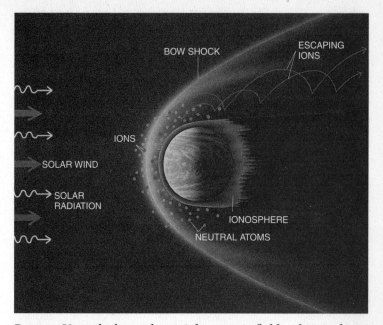

Because Venus lacks a substantial magnetic field, solar wind particles are far more interactive with the planet's atmosphere than they are with Earth's atmosphere. The Venusian ionosphere lies close to the planet's surface and forms a shell of permanently charged ions.

time of the *Pioneer Venus* mission, planetary scientists had accumulated strong evidence that the clouds were largely composed of concentrated solutions of sulfuric acid and water. They could not determine, however, the source of the sulfur from which the cloud droplets arose.

Pioneer Venus finally settled the question. As the *Orbiter* circled the planet, it scrutinized the tops of the clouds using its ultraviolet spectrometer, which identifies the characteristic pattern of emission and absorption from various atoms and molecules. Also, the gas chromatograph on the large probe measured the composition of the region below the main cloud deck. The results of these studies show that the sulfuric acid in the clouds derives from sulfur dioxide in the atmosphere.

Near the top of the clouds, some 60 to 70 kilometers above the ground, ultraviolet rays from the sun split sulfur dioxide into molecular fragments known as radicals. These radicals undergo a series of chemical reactions with radicals derived from water, ultimately producing tiny droplets of sulfuric acid. Gravity and air currents cause the droplets to migrate downward. As the droplets fall, they grow by colliding with one another and by accumulating sulfuric acid vapor from the air. At and below the base of the clouds, sulfuric acid particles dissociate back into sulfur dioxide and water vapor.

Instruments on the *Pioneer Venus* probes detected tiny particles (less than a thousandth of a millimeter across) at altitudes between 48 and 30 kilometers, just below the base of the cloud deck. Atmospheric motions carry these particles, along with sulfuric acid vapor, to higher, colder altitudes. There the sulfuric acid rapidly condenses onto the particles, producing much larger cloud particles that are concentrated toward the clouds' base. The density of those particles varies from place to place in the lower cloud region, probably because of irregularities in the up-welling and downwelling motions.

A related *Pioneer Venus* observation has stirred great excitement and controversy. In the course of exploring Venus's sulfur chemistry, the *Orbiter* detected an apparent, steady decrease in the concentration of sulfur dioxide near the tops of the clouds. Some workers interpreted that measurement as evidence that a giant volcanic eruption spewed sulfur into the atmosphere just about the time that *Pioneer Venus* arrived—a tantalizing sign that Venus might have active, explosive volcanoes. Once the eruption ceased, the sulfur levels would have begun to drop, as observed. Other investigators have argued that the changes in composition could have resulted from normal variations in the atmospheric circulation. The issue remains frustratingly unsettled.

Although it could not resolve the sulfur dioxide puzzle, *Pio-*

neer Venus has provided many other intriguing details about the circulation of the Venusian atmosphere. Such information is a tremendous boon for scientists attempting to understand atmospheric dynamics because it shows how weather patterns operate on a planet that differs from the earth in several crucial aspects.

Venus rotates extremely slowly; the earth completes 243 daily rotations in the time it takes Venus to turn once with respect to the stars. Also, because of the dense atmosphere, the surface temperature on Venus is nearly constant, from equator to pole. One might naively assume, therefore, that winds on Venus would be very sluggish.

Pioneer Venus proved that assumption false. On the earth, winds at low latitudes move more slowly than the rotation of the planet, whereas those at higher latitudes exceed the speed of the surface, a state known as superrotation. The atmosphere of Venus superrotates at all latitudes and at all heights from close to the surface to at least 90 kilometers above the surface. The winds attain their peak velocity near the cloud tops, where they blow at an unexpectedly rapid 100 meters a second, about 60 times as fast as the rotation of the underlying surface.

Winds in the atmospheres of the earth and the other terrestrial planets are driven by local imbalances between the amount of incoming solar energy and the amount of outgoing, radiated heat. In general, low latitudes, which receive the most sunlight, experience a net heating, whereas high latitudes, which receive the least incident solar energy, undergo a net cooling. As a result, the atmosphere develops a large-scale circulation pattern called a Hadley cell. In this pattern, hot air rises near the equator and travels toward the poles, where it sinks and returns toward the equator.

The spinning of a planet on its axis deflects the north-south (meridional) winds sideways, however, giving rise to east-west, or zonal, winds. Surprisingly, zonal winds almost always end up being much stronger than the north-south winds from which they derived. On the earth, Hadley circulation dominates atmospheric motions at low latitudes. Zonal winds close to the equator move slower than the earth's rotation (and hence are called easterlies); those closer to the poles form superrotating westerlies, culminating in the rapid flow of the jet stream. What is so odd about Venus's zonal winds is that they superrotate at almost all latitudes in the lower atmosphere.

Even now, planetary scientists do not fully understand why the entire lower atmosphere of Venus superrotates. The large fraction of solar energy that is absorbed high in the atmosphere, near the tops of the clouds, probably contributes to the brisk winds. The high-altitude heating of the atmosphere may set up a circulation system that is much less influenced by frictional interaction with the surface than is the case on the earth. The atmosphere of Venus might therefore be highly susceptible to the formation of eddies that can efficiently transport angular momentum. Such eddies could counteract the ability of the Hadley circulation to prevent superrotation at low latitudes. Cloud images taken by the *Pioneer Venus Orbiter* provide evidence of small-scale, eddylike variations in the winds.

High above the superrotating layers of the Venusian atmosphere lies the ionosphere, an extended zone of electrically charged atoms and molecules, or ions. The ions arise when high-energy ultraviolet rays from the sun knock electrons free from atmospheric gases. Every planet that has a substantial atmosphere possesses an ionosphere, but the one on Venus has a number of unusual traits.

The *Pioneer Venus Orbiter* monitored the passage of radio waves through the ionosphere and, during close approaches to the planet, measured its temperature, density and composition directly. As one might expect, Venus's ionosphere is densest in the center of the dayside hemisphere, near the equator, where the incoming sunlight is most direct. Because of the abundant chemical reactions occurring among the particles, Venus's ionosphere consists primarily of oxygen ions, even though carbon dioxide is the dominant gas at lower levels.

Unlike the earth and most other planets, Venus has no significant global magnetic field, for reasons not fully understood. The absence of a magnetic field significantly affects the structure of Venus's ionosphere. The *Orbiter* detected a weak ionosphere that extends beyond the day-night boundary. This finding was intriguing because, in the darkness, ions and free electrons should quickly recombine into neutral atoms. An instrument on the *Orbiter* found that on Venus ions from the dayside are able to migrate to the nightside. On the earth, the planetary magnetic field in the ionosphere inhibits such horizontal flow.

Images of the planet in ultraviolet radiation, obtained using the *Orbiter*'s ultraviolet spectrometer, detected a previously unknown, patchy aurora on Venus's shadowed hemisphere. Scientists attribute the aurora to energetic particles, probably fast-moving electrons, that crash into the atmosphere on the nightside. When these particles strike gas molecules in the atmosphere, they excite and ionize the molecules, further contributing to Venus's nighttime ionosphere. The excited molecules soon return to their normal, low-energy state by emitting radiation, which shows up as the aurora.

As is the case for terrestrial auroras, the particles that cause the auroras on Venus derive their energy from the solar wind. The solar wind is the sun's extended, rarefied outer atmosphere. It consists of plasma, or charged particles (primarily protons and electrons), racing from the sun at supersonic speeds. At the orbit of Venus the solar wind has a density of 15 protons and electrons per cubic centimeter and a velocity of 400 kilometers per second. As the solar wind blows past the planets, it carries part of the sun's magnetic field with it.

The intrinsic magnetic fields around the earth and other

planets act as obstacles to the electrically charged solar wind. The wind flows around those fields along a surface (the magnetopause) where the pressure of the wind equals the opposing magnetic pressure. The extent of the deflection depends on the strength of the planetary magnetic field. Venus, which has virtually no field at all, creates an obstacle scarcely bigger than the planet itself.

Nevertheless, the spacecraft found that solar-wind plasma was clearly being diverted around Venus. That discovery confirmed theoretical predictions that a planet's ionosphere can effectively block the solar wind even in the absence of a substantial magnetic field. Like a magnetic field, the ionosphere exerts pressure against the wind, but in this case it is the thermal pressure of the charged gas that counters the force of the solar wind. On average, the balance point lies at a 300-kilometer altitude near Venus's noontime equator and at 800 to 1,000 kilometers above the day-night boundary.

The deflection of the solar-wind flow around large obstacles (such as planets) is preceded by a "bow shock," a sharp boundary closely analogous to the shock that forms in front of a supersonic aircraft. During most of its lifetime, the *Pioneer Venus Orbiter* crossed the bow shock twice each lap, enabling it to monitor continuous changes in the magnetic environment around Venus. The craft found that the bow shock expands and contracts in step with the 11-year cycle of solar activity. The radius of the shock in the plane of the day-night boundary ranges in size from about 14,500 kilometers at solar maximum to 12,500 kilometers at solar minimum. The expansion and contraction probably result from changes in Venus's upper atmosphere associated with the varying radiation flux from the sun.

Just downstream of the bow shock, the solar wind grows more dense, slows down and changes its direction of flow. Magnetic-field lines are frozen into the solar wind because it is completely ionized. After the solar wind passes through the shock, the frozen-in interplanetary magnetic field piles up.

Pioneer Venus mapped the large-scale magnetic-field geometry around Venus. These data give the impression that the magnetic-field lines eventually slip around the obstacle and into the wake that it creates in the solar wind. Researchers refer to this wake structure as an induced magnetotail because it derives from the interplanetary magnetic field rather than from the planet's own field, as is the case for the earth's own, much larger magnetic tail.

Because of its lack of a significant internal field, Venus interacts more directly with the solar wind than does the earth. Over the age of the solar system, that interaction has affected the atmosphere of Venus. The planet's upper atmosphere, where atomic oxygen predominates, extends well above the point at which the solar wind is diverted around the planet. This gas remains largely unaffected by the solar-wind plasma as long as it remains electrically neutral. If an oxygen atom is struck by an ultraviolet ray or if it collides with a particle in the solar wind, it can become ionized. The oxygen ion couples to the flowing plasma, which may carry it away from the planet and out of the solar system.

Instruments on board the *Pioneer Venus Orbiter* confirm that the solar wind truly does scavenge Venus's upper atmosphere. Measurements of the density of the Venusian ionosphere indicate that the uppermost layers—those above the deduced height of the solar-wind obstacle—appear to be missing. Evidently the ions created above the obstacle have been carried off in the manner just described. The *Orbiter* has also detected the oxygen ions escaping tailward in the solar wind. In essence, *Pioneer Venus* has captured a snapshot of one of the processes by which Venus evolved into a world so unlike the earth.

Perhaps one of the most daring and most successful planetary exploration missions ever, the Mars Pathfinder *Mission provided a host of new data on the red planet. The mission's landing craft opened up to unleash a small rover that wandered the neighborhood of the landing site to study the local geology. Evidence from the mission suggests that Mars was once a wetter and warmer planet, probably sporting rivers, lakes, and perhaps even an ocean.*

The *Mars Pathfinder* Mission

Matthew P. Golombek

R ocks, rocks, look at those rocks," I exclaimed to everyone in the *Mars Pathfinder* control room at about 4:30 P.M. on July 4, 1997. The *Pathfinder* lander was sending back its first images of the surface of Mars, and everyone was focused on the television screens. We had gone to Mars to look at rocks, but no one knew for sure whether we would find any, because the landing site had been selected using orbital images with a resolution of roughly a kilometer. *Pathfinder* could have landed on a flat, rock-free plain. The first radio downlink indicated that the lander was nearly horizontal, which was worrisome for those of us interested in rocks, as most expected that a rocky surface would result in a tilted lander. The very first images were of the lander so that we could ascertain its condition, and it was not until a few tense minutes later that the first pictures of the surface showed a rocky plain—exactly as we had hoped and planned for.

Why did we want rocks? Every rock carries the history of its formation locked in its minerals, so we hoped the rocks would tell us about the early Martian environment. The two-part

Pathfinder payload, consisting of a main lander with a multi-spectral camera and a mobile rover with a chemical analyzer, was suited to looking at rocks. Although it could not identify the minerals directly—its analyzer could measure only their constituent chemical elements—our plan was to identify them indirectly based on the elemental composition and the shapes, textures and colors of the rocks. By landing *Pathfinder* at the mouth of a giant channel where a huge volume of water once flowed briefly, we sought rocks that had washed down from the ancient, heavily cratered highlands. Such rocks could offer clues to the early climate of Mars and to whether conditions were once conducive to the development of life.

The most important requirement for life on Earth (the only kind we know) is liquid water. Under present conditions on Mars, liquid water is unstable: because the temperature and pressure are so low, water is stable only as ice or vapor; liquid would survive for just a brief time before freezing or evaporating. Yet *Viking* images taken two decades ago show drainage channels and evidence for lakes in the highlands. These features hint at a warmer and wetter past on Mars in which water could persist on the surface. To be sure, other explanations have also been suggested, such as sapping processes driven by geothermal heating in an otherwise frigid and dry environment. One of *Pathfinder's* scientific goals was to look for evidence of a formerly warm, wet Mars.

The possible lake beds are found in terrain that, judging from its density of impact craters, is roughly the same age as the oldest rocks on Earth, which show clear evidence for life 3.9 billion to 3.6 billion years ago. If life was able to develop on Earth at this time, why not on Mars, too, if the conditions were similar? This is what makes studying Mars so compelling. By exploring our neighboring planet, we can seek answers to some of the most important questions in science: Are we alone in the universe? Will life arise anywhere that liquid water is stable, or does the formation of life require something else as well? And if life did develop on Mars, what happened to it? If life did not develop, why not?

Pathfinder was a Discovery-class mission—one of NASA's "faster, cheaper, better" spacecraft—to demonstrate a low-cost means of landing a small payload and mobile vehicle on Mars. It was developed, launched and operated under a fixed budget comparable to that of a major motion picture (between $200 million and $300 million), which is a mere fraction of the budget typically allocated for space missions. Built and launched in a short time (three and a half years), *Pathfinder* included three science instruments: the Imager for *Mars Pathfinder*, the Alpha Proton X-ray Spectrometer and the Atmospheric Structure Instrument/Meteorology Package. The rover itself also acted as an instrument; it was used to conduct 10 technology experiments, which studied the abrasion of metal films on a wheel of the rover and the adherence of dust to a solar cell as well as other ways the equipment on *Pathfinder* reacted to its surroundings.

In comparison, the *Viking* mission, which included two orbiter-lander pairs, was carried out more than 20 years ago at roughly 20 times the cost. *Viking* was very successful, returning more than 57,000 images that scientists have been studying ever since. The landers carried sophisticated experiments that tested for organisms at two locations; they found none.

The hardest part of *Pathfinder's* mission was the five minutes during which the spacecraft went from the relative security of interplanetary cruising to the stress of atmospheric entry, descent and landing. In that short time, more than 50 critical events had to be triggered at exactly the right times for the spacecraft to land safely. About 30 minutes before entry, the backpack-style cruise stage separated from the rest of the lander. At 130 kilometers above the surface, the spacecraft entered the atmosphere behind a protective aeroshell. A parachute unfurled 134 seconds before landing, and then the aeroshell was jettisoned. During descent, the lander was lowered beneath its back cover on a 20-meter-long bridle, or tether.

As *Pathfinder* approached the surface, its radar altimeter triggered the firing of three small solid-fuel rockets to slow it down further. Giant air bags inflated around each face of the tetrahedral lander, the bridle was cut, and the lander bounced onto the Martian surface at 50 kilometers per hour. Accelerometer measurements indicate that the air-bag-enshrouded lander bounced at least 15 times without losing air-bag pressure. After rolling at last to a stop, the lander deflated the air bags and opened to begin surface operations.

Although demonstrating this novel landing sequence was actually *Pathfinder's* primary goal, the rest of the mission also met or exceeded expectations. The lander lasted three times longer than its minimum design criteria, the rover 12 times longer. The mission returned 2.3 billion bits of new data from Mars, including more than 16,500 lander and 550 rover images and roughly 8.5 million individual temperature, pressure and wind measurements. The rover traversed a total of 100 meters in 230 commanded movements, thereby exploring more than 200 square meters of the surface. It obtained 16 measurements of rock and soil chemistry, performed soil-mechanics experiments and successfully completed the numerous technology experiments. The mission also captured the imagination of the public, garnering front-page headlines for a week and became the largest Internet event in history at the time, with a total of about 566 million hits for the first month of the mission—47 million on July 8, 1997, alone.

The mosaic of the landscape constructed from the first images revealed a rocky plain (about 20 percent of which was covered by rocks) that appears to have been deposited and shaped by catastrophic floods. This was what we had predicted based on remote-sensing data and the location of the landing site (19.13 degrees north, 33.22 degrees west), which is downstream from the mouth of Ares Vallis in the low area known as Chryse Plani-tia. In *Viking Orbiter* images, the area appears analogous to the Channeled Scabland in eastern and central Washington State. This analogy suggests that Ares Vallis formed when roughly the same volume of water as in the Great Lakes (hundreds of cubic kilometers) was catastrophically released, carving the observed channel in a few weeks. The density of impact craters in the region indicates it formed at an intermediate time in Mars's history, somewhere between 1.8 billion and 3.5 billion years ago.

The *Pathfinder* images support this interpretation. They show semirounded pebbles, cobbles and boulders similar to those deposited by terrestrial catastrophic floods. Rocks in what we dubbed the Rock Garden, a collection of rocks to the south-west of the lander, with the names Shark, Half Dome and Moe, are inclined and stacked, as if deposited by rapidly flowing water. Large rocks in the images (0.5 meter or larger) are flat-topped and often perched, also consistent with deposition by a flood. Twin Peaks, a pair of hills on the southwest horizon, are streamlined. *Viking* images suggest that the lander is on the flank of a broad, gentle ridge trending northeast from Twin Peaks; this ridge may be a debris tail deposited in the wake of the peaks. Small channels throughout the scene resemble those in the Channeled Scabland, where drainage in the last stage of the flood preferentially removed fine-grained materials.

The rocks in the scene are dark gray and covered with vari-ous amounts of yellowish-brown dust. This dust appears to be the same as that seen in the atmosphere, which, as imaging in different filters and locations in the sky suggests, is very fine grained (a few microns in diameter). The dust also collected in wind streaks behind rocks.

Some of the rocks have been fluted and grooved, presum-ably by sand-size particles (less than one millimeter) that hopped along the surface in the wind. The rover's camera also saw sand dunes in the trough behind the Rock Garden. Dirt covers the lower few centimeters of some rocks, suggesting that they have been exhumed by wind. Despite these signs of slow erosion by the wind, the rocks and surface appear to have changed little since they were deposited by the flood.

The Alpha Proton X-ray Spectrometer on the rover measured the compositions of eight rocks. The silicon content of some of the rocks is much higher than that of the Martian meteorites, our only other samples of Mars. The Martian meteorites are all maficigneous rocks, volcanic rocks that are relatively low in silicon and high in iron and magnesium. Such rocks form when the upper mantle of a planet melts. The melt rises up through the crust and solidifies at or near the surface. These types of rocks, referred to as basalts, are the most common rock on Earth and have also been found on the moon. Based on the composition of the Martian meteorites and the presence of plains and mountains that look like features produced by basaltic volcanism on Earth, geologists expected to find basalts on Mars.

The rocks analyzed by *Pathfinder*, however, are not basalts. If they are volcanic, as suggested by their vesicular surface texture, presumably formed when gases trapped during cooling left small holes in the rock, their silicon content classifies them as andesites. Andesites form when the basaltic melt from the mantle intrudes deep within the crust. Crystals rich in iron and magnesium form and sink back down, leaving a more silicon-rich melt that erupts onto the surface. The andesites were a great surprise, but because we do not know where these rocks came from on the Martian surface, we do not know the full implications of this discovery. If the andesites are representative of the highlands, they suggest that ancient crust on Mars is similar in composition to continental crust on Earth. This similarity would be difficult to reconcile with the very different geologic histories of the two planets. Alternatively, the rocks could represent a minor proportion of high-silicon rocks from a predominately basaltic plain.

Intriguingly, not all the rocks appear to be volcanic, judging by the diversity of morphologies, textures and fabrics observed in high-resolution images. Some rocks appear similar to impact breccias, which are composed of angular fragments of different materials. Others have layers like those in terrestrial sedimentary rocks, which form by deposition of smaller fragments of rocks in water. Indeed, rover images show many rounded pebbles and cobbles on the ground. In addition, some larger rocks have what look like embedded pebbles and shiny indentations, where it looks as though rounded pebbles that were pressed into the rock during its formation have fallen out, leaving holes. These rocks may be conglomerates formed by flowing liquid water. The water would have rounded the pebbles and deposited them in a sand, silt and clay matrix; the matrix was subsequently compressed, forming a rock, and carried to its present location by the flood. Because conglomerates require a long time to form, if these Martian rocks are conglomerates (other interpretations are also possible) they strongly suggest that liquid water was once stable on the planet and that the climate was therefore warmer and wetter than at present.

Soils at the landing site vary from bright reddish dust to darker-red and darker-gray material, generally consistent with fine-grained iron oxides. Overall, the soils are lower in silicon than the rocks and richer in sulfur, iron and magnesium. Soil compositions are generally similar to those measured at the *Viking* sites, which are on opposite hemispheres (*Viking 1* is 800 kilometers west of *Pathfinder*; *Viking 2* is thousands of kilometers away on the opposite, eastern side of the northern hemisphere). Thus, the soil appears to include materials distributed globally on Mars, such as the airborne dust. The similarity in compositions among the soils implies that the variations in color at each site may be the result of slight differences in iron mineralogy or in particle size and shape.

A bright reddish or pink material also covered part of the

site. Similar to the soils in composition, it seems to be indurated or cemented because it was not damaged by scraping with the rover wheels.

Pathfinder also investigated the dust in the atmosphere of Mars by observing its deposition on a series of magnetic targets on the spacecraft. The dust, it turned out, is highly magnetic. It may consist of small silicate (perhaps clay) particles, with some stain or cement of a highly magnetic mineral known as maghemite. This finding, too, is consistent with a watery past. The iron may have dissolved out of crustal materials in water, and the maghemite may be a freeze-dried precipitate.

The sky on Mars had the same butter-scotch color as it did when imaged by the *Viking* landers. Fine-grained dust in the atmosphere would explain this color. Hubble Space Telescope images had suggested a very clear atmosphere; scientists thought it might even appear blue from the surface. But *Pathfinder* found otherwise, suggesting either that the atmosphere always has some dust in it from local dust storms or dust devils, or that the atmospheric opacity varies appreciably over a short time. The inferred dust-particle shape and size (a few microns in diameter) and the amount of water vapor in the atmosphere (equivalent to a pitiful hundredth of a millimeter of rainfall) are also consistent with measurements made by *Viking*. Even if Mars was once lush, it is now drier and dustier than any desert on Earth.

The meteorological sensors gave further information about the atmosphere. They found patterns of diurnal and longer-term pressure and temperature fluctuations. The temperature reached its maximum of 263 kelvins (−10 degrees Celsius) every day at 2:00 P.M. local solar time and its minimum of 197 kelvins (−76 degrees C), just before sunrise. The pressure minimum of just under 6.7 millibars (roughly 0.67 percent of pressure at sea level on Earth) was reached on sol 21, the 21st Martian day after landing. On Mars the air pressure varies with the seasons. During winter, it is so cold that 20 to 30 percent of the entire atmosphere freezes out at the pole, forming a huge pile of solid carbon dioxide. The pressure minimum seen by *Pathfinder* indicates that the atmosphere was at its thinnest, and the south polar cap its largest, on sol 21.

Morning temperatures fluctuated abruptly with time and height; the sensors positioned 0.25, 0.5 and one meter above the spacecraft took different readings. If you were standing on Mars, your nose would be at least 20 degrees C colder than your feet. This suggests that cold morning air is warmed by the surface and rises in small eddies, or whirlpools, which is very different from what happens on Earth, where such large temperature disparities do not occur. Afternoon temperatures, after the air has warmed, do not show these variations.

In the early afternoon, dust devils repeatedly swept across the lander. They showed up as sharp, short-lived pressure changes with rapid shifts in wind direction; they also appear in images as dusty funnel-shaped vortices tens of meters across and hundreds of meters high. They were probably similar to events detected by the *Viking* landers and orbiters and may be an important mechanism for raising dust into the Martian atmosphere. Otherwise, the prevailing winds were light (clocked at less than 36 kilometers per hour) and variable.

Pathfinder measured atmospheric conditions at higher altitudes during its descent. The upper atmosphere (altitude above 60 kilometers) was colder than *Viking* had measured. This finding may simply reflect seasonal variations and the time of entry: *Pathfinder* came in at 3:00 A.M. local solar time, whereas *Viking* arrived at 4:00 P.M., when the atmosphere is naturally warmer. The lower atmosphere was similar to that measured by *Viking*, and its conditions can be attributed to dust mixed uniformly in comparatively warm air.

As a bonus, mission scientists were able to use radio communications signals from *Pathfinder* to measure the rotation of Mars. Daily Doppler tracking and less frequent two-way ranging during communication sessions determined the position of the lander with a precision of 100 meters. The last such positional measurement was done by *Viking* more than 20 years ago. In the

Summary of Evidence for a Warmer, Wetter Mars

Over the past three decades, scientists have built the case that Mars once looked much like Earth, with rainfall, rivers, lakes, maybe even an ocean. *Pathfinder* has added evidence that strengthens this case.

GEOLOGIC FEATURE	PROBABLE ORIGIN	IMPLICATION
Riverlike valley networks	Water flow out of ground or from rain	Either atmosphere was thicker (allowing rain) or geothermal heating was stronger (causing groundwater sapping)
Central channel ("thalweg") in broader valleys	Fluid flow down valley center	Valleys were formed by water flow, not by land slides or sapping
Lakelike depressions with drainage networks; layered deposits in canyons	Flow through channels into lake	Water existed at the surface, but for unknown time
Possible strand lines and erosional beaches and terraces	Possible shoreline	Northern hemisphere might have had an ocean
Rimless craters and highly eroded ancient terrain	High erosion rates	Water, including rain, eroded surface
Rounded pebbles and possible conglomerate rock	Rock formation in flowing water	Liquid water was stable, so atmosphere was thicker and warmer
Abundant sand	Action of water on rocks	Water was widespread
Highly magnetic dust	Maghemite stain or cement on small (micron-size) silicate grains	Active hydrologic cycle leached iron from crustal materials to form maghemite

interim, the pole of rotation has precessed—that is, the direction of the tilt of the planet has changed, just as a spinning top slowly wobbles. The difference between the two positional measurements yields the precession rate. The rate is governed by the moment of inertia of the planet, a function of the distribution of mass within the planet. The moment of inertia had been the most important number about Mars that we did not yet know.

From *Pathfinder's* determination of the moment of inertia we now know that Mars must have a central metallic core that is between 1,300 and 2,400 kilometers in radius. With assumptions about the mantle composition, derived from the compositions of the Martian meteorites and the rocks measured by the rover, scientists can now start to put constraints on interior temperatures. Before *Pathfinder*, the composition of the Martian meteorites argued for a core, but the size of this core was completely unknown. The new information about the interior will help geophysicists understand how Mars has evolved over time.

Taking all the results together suggests that Mars was once more Earthlike than previously appreciated. Some crustal materials on Mars resemble, in silicon content, continental crust on Earth. Moreover, the rounded pebbles and the possible conglomerate, as well as the abundant sand-and dust-size particles, argue for a formerly water-rich planet. The earlier environment may have been warmer and wetter, perhaps similar to that of the early Earth. In contrast, since floods produced the landing site 1.8 billion to 3.5 billion years ago, Mars has been a very un-Earthlike place. The site appears almost unaltered since it was deposited, indicating very low erosion rates and thus no water in relatively recent times.

Although we are not certain that early Mars was more like Earth, the data returned from *Pathfinder* are very suggestive. Information from the *Mars Global Surveyor*, now orbiting the Red Planet, should help answer this crucial question about our neighboring world.

The present-day orbits of the planets in our solar system are probably the results of millions and millions of years of slow evolution. New evidence points toward a very different orbital pattern for the planets shortly after the solar system's birth. The planetary orbits we observe today are thought to be the result of a complex interaction of the gravity of the planets with one another and with the sun.

Migrating Planets

Renu Malhotra

I n the familiar renditions of the solar system, each planet moves around the sun in its own well-defined orbit, maintaining a respectful distance from its neighbors. The planets have maintained this celestial merry-go-round since astronomers began recording their motions, and mathematical models show that this stable orbital configuration has existed for almost the entire 4.5-billion-year history of the solar system. It is tempting, then, to assume that the planets were "born" in the orbits that we now observe.

Certainly it is the simplest hypothesis. Modern-day astronomers have generally presumed that the observed distances of the planets from the sun indicate their birthplaces in the solar nebula, the primordial disk of dust and gas that gave rise to the solar system. The orbital radii of the planets have been used to infer the mass distribution within the solar nebula. With this basic information, theorists have derived constraints on the nature and timescales of planetary formation. Consequently, much of our understanding of the early history

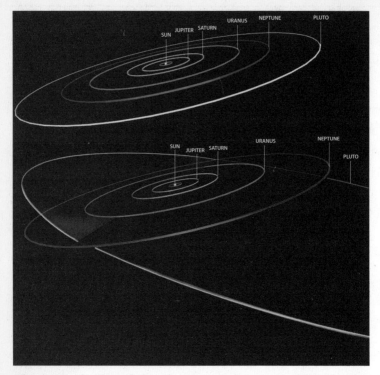

The original orbit of the planets may have resembled the diagram on the top. Over time, the arrangement shifted: Jupiter's orbit shrunk slightly, while the orbits of Saturn, Uranus, and Neptune expanded. Neptune's gravity caused Pluto's orbit to become more eccentric.

of the solar system is based on the assumption that the planets formed in their current orbits.

It is widely accepted, however, that many of the smaller bodies in the solar system—asteroids, comets and the planets' moons—have altered their orbits over the past 4.5 billion years, some more dramatically than others. The demise of Comet Shoemaker-Levy 9 when it collided with Jupiter in 1994 was striking evidence of the dynamic nature of some objects in the solar system. Still smaller objects—micron- and millimeter-size interplanetary particles shaken loose from comets and asteroids—undergo a more gradual orbital evolution, gently spiraling in toward the sun and raining down on the planets in their path.

Furthermore, the orbits of many planetary satellites have changed significantly since their formation. For example, Earth's moon is believed to have formed within 30,000 kilometers (18,600 miles) of Earth—but it now orbits at a distance of 384,000 kilometers. The moon has receded by nearly 100,000 kilometers in just the past billion years because of tidal forces (small gravitational torques) exerted by our planet. Also, many satellites of the outer planets orbit in lockstep with one another: for instance, the orbital period of Ganymede, Jupiter's largest moon, is twice that of Europa, which in turn has a period twice that of Io. This precise synchronization is believed to be the result of a gradual evolution of the satellites' orbits by means of tidal forces exerted by the planet they are circling.

Until recently, little provoked the idea that the orbital configuration of the planets has altered significantly since their formation. But some remarkable developments during the past five years indicate that the planets may indeed have migrated from their original orbits. The discovery of the Kuiper belt has shown that our solar system does not end at Pluto (see "The Kuiper Belt," page 111). Approximately 100,000 icy "minor planets" (ranging between 100 and 1,000 kilometers in diameter) and an even greater number of smaller bodies occupy a region extending from Neptune's orbit—about 4.5 billion kilometers from the sun—to at least twice that distance. The distribution of these objects exhibits prominent nonrandom features that cannot be readily explained by the current model of the solar system. Theoretical models for the origin of these peculiarities suggest the intriguing possibility that the Kuiper belt bears traces of the orbital history of the gas-giant planets—specifically, evidence of a slow spreading of these planets' orbits subsequent to their formation.

What is more, the recent discovery of several Jupiter-size companions orbiting nearby sunlike stars in peculiarly small orbits has also focused attention on planetary migration. It is difficult to understand the formation of these putative planets at such small distances from their parent stars. Hypotheses for their origin have proposed that they accreted at more comfortable distances from their parent stars—similar to the distance between Jupiter and the sun—and then migrated to their present positions.

Until just a few years ago, the only planetary objects known beyond Neptune were Pluto and its satellite, Charon. Pluto has long been a misfit in the prevailing theories of the solar system's origin: it is thousands of times less massive than the four gas-giant outer planets, and its orbit is very different from the well-separated, nearly circular and co-planar orbits of the eight other major planets. Pluto's is eccentric: during one complete revolution, the planet's distance from the sun varies from 29.7 to 49.5 astronomical units (one astronomical unit, or AU, is the distance between Earth and the sun, about 150 million kilometers). Pluto also travels 8 AU above and 13 AU below the mean plane of the other planets' orbits. For approximately two decades in its orbital period of 248 years, Pluto is closer to the sun than Neptune is.

In the decades since Pluto's discovery in 1930, the planet's enigma has deepened. Astronomers have found that most Neptune-crossing orbits are unstable—a body in such an orbit will either collide with Neptune or be ejected from the outer solar system in a relatively short time, typically less than 1 percent of the age of the solar system. But the particular Neptune-crossing orbit in which Pluto travels is protected from close approaches to the gas giant by a phenomenon called resonance libration. Pluto makes two revolutions around the sun during the time that Neptune makes three; Pluto's orbit is therefore said to be in 3:2 resonance with Neptune's. The relative motions of the two planets ensure that when Pluto crosses Neptune's orbit, it is far away from the larger planet.

In addition, Pluto's perihelion—its closest approach to the sun—always occurs high above the plane of Neptune's orbit, thus maintaining Pluto's long-term orbital stability. Computer simulations of the orbital motions of the outer planets, including the effects of their mutual perturbations, indicate that the relationship between the orbits of Pluto and Neptune is billions of years old and will persist for billions of years.

How did Pluto come to have such a peculiar orbit? In the past, this question has stimulated several speculative and ad hoc explanations, typically involving planetary encounters. Recently, however, significant advances have been made in understanding the complex dynamics of orbital resonances and in identifying their Jekyll-and-Hyde role in producing both chaos and exceptional stability in the solar system. Drawing on this body of knowledge, I proposed in 1993 that Pluto was born somewhat beyond Neptune and initially traveled in a nearly circular, low-inclination orbit similar to those of the other planets but that it was transported to its current orbit by resonant gravitational interactions with Neptune. A key feature of this theory is that it abandons the assumption that the gas-giant planets formed at their present distances from the sun. Instead it proposes an epoch of planetary orbital migration early in the history of the solar system, with Pluto's unusual orbit as evidence of that migration.

The story begins at a stage when the process of planetary formation was almost but not quite complete. The gas giants— Jupiter, Saturn, Uranus and Neptune—had nearly finished coalescing from the solar nebula, but a residual population of small planetesimals—rocky and icy bodies, most no larger than a few tens of kilometers in diameter—remained in their midst. The relatively slower subsequent evolution of the solar system consisted of the scattering or accretion of the planetesimals by the major planets. Because the planetary scattering ejected most of the planetesimal debris to distant or unbound orbits—essentially throwing the bodies out of the solar system—there was a net loss of orbital energy and angular momentum from the giant planets' orbits. But because of their different masses and distances from the sun, this loss was not evenly shared by the four giant planets.

In particular, consider the orbital evolution of the outermost giant planet, Neptune, as it scattered the swarm of planetesimals in its vicinity. At first, the mean specific orbital energy of the planetesimals (the orbital energy per unit of mass) was equal to that of Neptune itself, so Neptune did not gain or lose energy from its gravitational interactions with the bodies. At later times, however, the planetesimal swarm near Neptune was depleted of the lower-energy objects, which had moved into the gravitational reach of the other giant planets. Most of these planetesimals were eventually ejected from the solar system by Jupiter, the heavyweight of the planets.

Thus, as time went on, the specific orbital energy of the planetesimals that Neptune encountered grew larger than that of Neptune itself. During subsequent scatterings, Neptune gained orbital energy and migrated outward. Saturn and Uranus also gained orbital energy and spiraled outward. In contrast, Jupiter lost orbital energy; its loss balanced the gains of the other planets and planetesimals, hence conserving the total energy of the system. But because Jupiter is so massive and had so much orbital energy and angular momentum to begin with, its orbit decayed only slightly.

The possibility of such subtle adjustments of the giant planets' orbits was first described in a little-noticed paper published in 1984 by Julio A. Fernandez and Wing-Huen Ip, a Uruguayan and Taiwanese astronomer duo working at the Max Planck Institute in Germany. Their work remained a curiosity and escaped any comment among planet formation theorists, possibly because no supporting observations or theoretical consequences had been identified.

In 1993 I theorized that as Neptune's orbit slowly expanded, the orbits that would be resonant with Neptune's also expanded. In fact, these resonant orbits would have swept by Pluto, assuming that the planet was originally in a nearly circular, low-inclination orbit beyond Neptune. I calculated that any such objects would have had a high probability of being "captured" and pushed outward along the resonant orbits as Neptune migrated. As these bodies moved outward, their orbital eccentricities and inclinations would have been driven to larger values by the resonant gravitational torque from Neptune. (This effect is analogous to the pumping-up of the amplitude of a playground swing by means of small periodic pushes at the swing's natural frequency.) The final maximum eccentricity would therefore provide a direct measure of the magnitude of Neptune's migration. According to this theory, Pluto's orbital eccentricity of 0.25 suggests that Neptune has migrated outward by at least 5 AU. Later, with the help of computer simulations, I revised this to 8 AU and also estimated that the timescale of migration had to be a few tens of millions of years to account for the inclination of Pluto's orbit.

Of course, if Pluto were the only object beyond Neptune, this explanation of its orbit, though compelling in many of its details, would have remained unverifiable. The theory makes specific predictions, however, about the orbital distribution of bodies in the Kuiper belt, which is the remnant of the primordial disk of planetesimals beyond Neptune. Provided that the largest bodies in the primordial Kuiper belt were sufficiently small that their perturbations on the other objects in the belt would be negligible, the dynamical mechanism of resonance sweeping would work not only on Pluto but on all the trans-Neptunian objects, perturbing them from their original orbits. As a result, prominent concentrations of objects in eccentric orbits would be found at Neptune's two strongest resonances, the 3:2 and the 2:1. Such orbits are ellipses with semimajor axes of 39.5 AU and 47.8 AU, respectively. (The length of the semimajor axis is equal to the object's average distance from the sun.)

More modest concentrations of trans-Neptunian bodies would be found at other resonances, such as the 5:3. The population of objects closer to Neptune than the 3:2 resonant orbit would be severely depleted because of the thorough resonance sweeping of that region and because perturbations caused by Neptune would destabilize the orbits of any bodies that remained. On the other hand, planetesimals that accreted beyond 50 AU from the sun would be expected to be largely unperturbed and still orbiting in their primordial distribution.

Fortunately, recent observations of Kuiper belt objects, or KBOs, have provided a means of testing this theory. More than 174 KBOs have been discovered as of mid-1999. Most have orbital periods in excess of 250 years and thus have been tracked for less than 1 percent of their orbits. Nevertheless, reasonably reliable orbital parameters have been determined for about 45 of the known KBOs. Their orbital distribution is not a pattern of uniform, nearly circular, low-inclination orbits, as would be expected for a pristine, unperturbed planetesimal population. Instead one finds strong evidence of gaps and concentrations in the distribution. A large fraction of these KBOs travel in eccentric 3:2 resonant orbits similar to Pluto's, and KBOs in orbits interior to the 3:2 orbit are nearly absent— which is consistent with the predictions of the resonance sweeping theory.

Still, one outstanding question remains: Are there KBOs in the 2:1 resonance comparable in number to those found in the 3:2, as the planet migration theory would suggest? And what is the orbital distribution at even greater distances from the sun? At present, the census of the Kuiper belt is too incomplete to answer this question fully. But on Christmas Eve 1998 the Minor Planet Center in Cambridge, Mass., announced the identification of the first KBO orbiting in 2:1 resonance with

Neptune. Two days later the center revealed that another KBO was traveling in a 2:1 resonant orbit. Both these objects have large orbital eccentricities, and they may turn out to be members of a substantial population of KBOs in similar orbits. They had previously been identified as orbiting in the 3:2 and 5:3 resonances, respectively, but new observations made last year strongly indicated that the original identifications were incorrect. This episode underscored the need for continued tracking of known KBOs in order to map their orbital distribution correctly. We must also acknowledge the dangers of overinterpreting a still small data set of KBO orbits.

In short, although other explanations cannot be ruled out yet, the orbital distribution of KBOs provides increasingly strong evidence for planetary migration. The data suggest that Neptune was born about 3.3 billion kilometers from the sun and then moved about 1.2 billion kilometers outward—a journey of almost 30 percent of its present orbital radius. For Uranus, Saturn and Jupiter, the magnitude of migration was smaller, perhaps 15, 10 and 2 percent, respectively; the estimates are less certain for these planets because, unlike Neptune, they could not leave a direct imprint on the Kuiper belt population.

Most of this migration took place over a period shorter than 100 million years. That is long compared with the timescale for the formation of the planets—which most likely took less than 10 million years—but short compared with the 4.5-billion-year age of the solar system. In other words, the planetary migration occurred in the early history of the solar system but during the later stages of planet formation. The total mass of the scattered planetesimals was about three times Neptune's mass. The question arises whether even more drastic orbital changes might occur in planetary systems at earlier times, when the primordial disk of dust and gas contains more matter and perhaps many protoplanets in nearby orbits competing in the accretion process.

* * *

In the early 1980s theoretical studies by Peter Goldreich and Scott Tremaine, both then at the California Institute of Technology, and others concluded that the gravitational forces between a protoplanet and the surrounding disk of gas, as well as the energy losses caused by viscous forces in a gaseous medium, could lead to very large exchanges of energy and angular momentum between the protoplanet and the disk. If the torques exerted on the protoplanet by the disk matter just inside the planet's orbit and by the matter just beyond it were slightly unbalanced, rapid and drastic changes in the planet's orbit could happen. But again, this theoretical possibility received little attention from other astronomers at the time. Having only our solar system as an example, planet formation theorists continued to assume that the planets were born in their currently observed orbits.

In the past five years, however, the search for extrasolar planets has yielded possible signs of planetary migration. By measuring the telltale wobbles of nearby stars—within 50 light-years of our solar system—astronomers have found evidence of more than a dozen Jupiter-mass companions in surprisingly small orbits around main-sequence stars. The first putative planet was detected orbiting the star 51 Pegasi in 1995 by two Swiss astronomers, Michel Mayor and Didier Queloz of the Geneva Observatory, who were actually surveying for binary stars. Their observations were quickly confirmed by Geoffrey W. Marcy and R. Paul Butler, two American astronomers working at Lick Observatory near San Jose, Calif. As of June 1999, 20 extrasolar planetary candidates have been identified, most by Marcy and Butler, in search programs that have surveyed almost 500 nearby sunlike stars over the past 10 years. The technique used in these searches—measuring the Doppler shifts in the stars' spectral lines to determine periodic variations in stellar velocities—yields only a lower limit on the masses of the stars' companions. Most of the candidate planets have minimum masses of about one Jupiter-mass and orbital radii shorter than 0.5 AU.

What is the relationship between these objects and the planets in our solar system? According to the prevailing model of planet formation, the giant planets in our solar system coalesced in a two-step process. In the first step, solid planetesimals clumped together to form a proto-planetary core. Then this core gravitationally attracted a massive gaseous envelope from the surrounding nebula. This process must have been completed within about 10 million years of the formation of the solar nebula itself, as inferred from astronomical observations of the lifetime of protoplanetary disks around young sunlike stars.

At distances of less than 0.5 AU from a star, there is insufficient mass in the primordial disk for solid protoplanetary cores to condense. Furthermore, it is questionable whether a protoplanet in a close orbit could attract enough ambient gas to provide the massive envelope of a Jupiterlike planet. One reason is simple geometry: an object in a tight orbit travels through a smaller volume of space than one in a large orbit does. Also, the gas disk is hotter close to the star and hence less likely to condense onto a protoplanetary core. These considerations have argued against the formation of giant planets in very short-period orbits.

Instead several theorists have suggested that the putative extrasolar giant planets may have formed at distances of several AU from the star and subsequently migrated inward. Three mechanisms for planetary orbital migration are under discussion. Two involve disk-protoplanet interactions that allow planets to move long distances from their birthplaces as long as a massive disk remains.

With the disk-protoplanet interactions theorized by Goldreich and Tremaine, the planet would be virtually locked to the inward flow of gas accreting onto the protostar and might either plunge into the star or decouple from the gas when it drew close to the star. The second mechanism is interaction with a planetesimal disk rather than a gas disk: a giant planet embedded in a very massive planetesimal disk would exchange energy and angular momentum with the disk through gravitational scattering and resonant interactions, and its orbit would shrink all the way to the disk's inner edge, just a few stellar radii from the star.

The third mechanism is the scattering of large planets that either formed in or moved into orbits too close to one another for long-term stability. In this process, the outcomes would be quite unpredictable but generally would yield very eccentric orbits for both planets. In some fortuitous cases, one of the scattered planets would move to an eccentric orbit that would come so near the star at its closest approach that tidal friction would eventually circularize its orbit; the other planet, meanwhile, would be scattered to a distant eccentric orbit. All the mechanisms accommodate a broad range of final orbital radii and orbital eccentricities for the surviving planets.

These ideas are more than a simple tweak of the standard model of planet formation. They challenge the widely held expectation that protoplanetary disks around sunlike stars commonly evolve into regular planetary systems like our own. It is possible that most planets are born in unstable configurations and that subsequent planet migration can lead to quite different results in each system, depending sensitively on initial disk properties. An elucidation of the relation between the newly discovered extrasolar companions and the planets in our solar system awaits further theoretical and observational developments. Nevertheless, one thing is certain: the idea that planets can change their orbits dramatically is here to stay.

Long delayed and continually hampered by technical difficulties, when the Galileo Mission to Jupiter finally arrived at the solar system's largest planet, the wealth of new information garnered about the gas giant and its moons astounded space scientists. With several geologically-active moons that rival the planet Mercury in size, Jupiter and its moons are now recognized as a kind of "mini-solar system" whose complexity and subtleties rival Sol's planetary family.

The *Galileo* Mission to Jupiter and Its Moons

Torrence V. Johnson

To conserve power, the probe was traveling in radio silence, with only a small clock counting down the seconds. Racing 215,000 kilometers overhead, its companion spacecraft was ready to receive its transmissions. Back on Earth, engineers and scientists, many of whom had spent most of two decades involved in the project, awaited two key signals. The first was a single data bit, a simple yes or no indicating whether the little probe had survived its fiery plunge into Jupiter's massive atmosphere.

Getting this far had not been easy for the *Galileo* mission. When conceived in the mid-1970s, the two-part unmanned spacecraft was supposed to set forth in 1982, carried into Earth orbit on board the space shuttle and sent onward to Jupiter by a special upper rocket stage. But slips in the first shuttle launches and problems with upper-stage development kept pushing the schedule back. Then came the *Challenger* tragedy in 1986, which occurred just as *Galileo* was being readied for launch. Forced by the circumstances to switch to a safer but weaker upper stage, engineers had to plot a harrowing gravity-assist trajectory, using close flybys of Venus and Earth to provide the boost the new rocket could not. From launch in October 1989, the journey took six years. Two years into the flight, disaster struck again when the umbrellalike main communications antenna refused to unfurl, leaving the spacecraft with only its low-capacity backup antenna. Later, the tape recorder—vital for storing data—got stuck.

When engineers received the "golden bit" confirming that the probe was still alive, cheers went up in the control room and the tension began to ease. But the team still had to wait out the next two hours for the second critical event: insertion of the companion spacecraft into orbit. To slow it from interplanetary cruise enough for Jupiter's gravity to capture it, engineers instructed the German-built main engine to fire for 45 minutes. Finally, word came through that this maneuver had succeeded. The orbiter had become the first known artificial satellite of the giant planet.

The atmospheric probe penetrated the kaleidoscopic clouds and conducted the first in situ sampling of an outer planet's atmosphere, transmitting data for an hour before it was lost in the gaseous depths. The orbiter is still going strong. It has photographed and analyzed the planet, its rings and its diverse moons. Most famously, it has bolstered the case that an ocean of liquid water lurks inside Europa, one of the four natural satellites discovered by Galileo Galilei in 1610. But the other large moons have revealed surprises of their own: beams of electrons that connect Io, the most volcanically tormented body in the solar system, to Jupiter; a magnetic field generated within Ganymede, the first such field ever discovered on a moon; and the subtle mysteries of Callisto, including signs that it, too, has an ocean.

According to modern theories of planet formation, Jupiter and the other giant planets emerged from the primordial solar nebula in two stages. First, icy planetesimals—essentially large comets that had condensed out of the cloud of gas and dust—clumped together. Then, as the protoplanet grew to a certain critical size, it swept up gas directly from the nebula. Jupiter thus started off with a sample of the raw material of the solar system, which had roughly the same composition as the early sun. Since then, the planet has been shaped by processes such as internal differentiation and the continuing infall of cometary material. Disentangling these processes was the main goal of the atmospheric probe.

Perhaps the most mysterious discovery by the probe involved the so-called condensable species, including elements such as nitrogen, sulfur, oxygen and carbon. Scientists have long known that Jupiter has about three times as much carbon (in the form of methane gas) as the sun. The other species (in the form of ammonia, ammonia sulfides and water) are thought to condense and form cloud layers at various depths. Impurities in the cloud droplets, possibly sulfur or phosphorus, give each layer a distinctive color. The probe was designed to descend below the lowest expected cloud deck, believed to be a water cloud at about 5 to 10 atmospheres of pressure—some 100 kilometers below the upper ammonia ice clouds. The expected weather report was windy, cloudy, hot and humid.

Yet the instruments saw almost no evidence for clouds, detecting only light hazes at a pressure level of 1.6 atmospheres. The water and sulfur abundances were low. The lightning detector—basically an AM radio that listened for bursts of static—registered only faint discharges. In short, the weather was clear and dry. So what had gone wrong with the prediction? One piece of the answer came quickly. Infrared images from Earth-based telescopes discovered that the probe had unwittingly hit a special type of atmospheric region known as a five-micron hot spot—a clearing where infrared radiation from lower, hotter levels leaks out. Scientists had expected that even in these regions the gases at the depths the probe reached would match the average composition of the whole atmosphere. If so, Jupiter has an anomalously low amount of such elements as oxygen and sulfur. But no one has proposed a process that would eliminate these elements so efficiently. The other possibility is that the composition of the hot spot differs from the average, perhaps because of a massive downdraft of cold, dry gas from the upper atmosphere.

The latter theory has its own difficulties but currently seems the more likely interpretation. Just before the probe ceased transmitting, concentrations of water, ammonia and hydrogen sulfide were beginning to rise rapidly—just as if the probe was approaching the base of a downdraft. *Orbiter* images of another prominent hot spot show that winds converge on the center of the hot spot from all directions. The only place the gas can go is down. *Orbiter* spectra showed that the abundance of water and ammonia varies by a factor of 100 among different hot spots, supporting the idea that local meteorological conditions dictate the detailed composition of the atmosphere.

The one part of the weather prediction that proved correct was "windy." Jupiter's cloud bands are associated with high-velocity jet streams: westerlies and easterlies that blow steadily at several hundred kilometers per hour. On Earth the analogous winds die off near the surface. On Jupiter there is no surface; the wind profile depends on which energy source dominates the atmosphere. If a source of internal energy (such as slow contraction under the force of gravity) dominates, the winds should stay strong or increase with depth. The opposite is true if external energy (such as sunlight) is the main contributor. By tracking the probe's radio signal, scientists ascertained that winds at first increase rapidly with depth and then remain constant—indicating that Jupiter's atmosphere is driven by internal energy.

Although the probe detected only weak hints of lightning, the orbiter saw bright flashes illuminating the clouds in what are obviously massive thunderstorms. Like *Voyager*, *Galileo* found that lightning was concentrated in just a few zones of latitude. These zones are regions of anti-cyclonic shear: the winds change speed abruptly going from north to south, creating turbulent, stormy conditions. As on Earth, lightning may occur in water clouds where partially frozen ice granules rise and fall in the turbulence, causing positive and negative charges to separate. How deep the lightning occurs can be estimated from the size of the illuminated spot on the clouds; the bigger the spot, the deeper the discharge. *Galileo* deduced that the lightning is indeed originating from layers in the atmosphere where water clouds are expected to form.

For all its pains, the probe descended less than 0.1 percent of the way to the center of the planet before succumbing to the high pressures and temperatures. Nevertheless, some of its measurements hint at what happens deeper down. The concentrations of noble gases—helium (the second most abundant element in Jupiter, after hydrogen), neon, argon, krypton and xenon—are particularly instructive. Because these gases do not react chemically with other elements, they are comparatively unambiguous tracers of physical conditions within the planet. So informative is the concentration of helium that the *Galileo* atmospheric probe carried an instrument dedicated solely to its measurement.

Infrared spectra obtained by *Voyager* suggested that Jupiter contains proportionately much less helium than the sun does, an indication that something must have drained this element from the upper atmosphere. *Galileo*, however, found that Jupiter has nearly the same helium content as the outer layers of the sun. This result still requires that some process remove helium from the Jovian atmosphere, because the outer layers of the sun have themselves lost helium. But that process must have started later in the planet's history than researchers had thought. *Galileo* also discovered that the concentration of neon is a tenth of its solar value.

Both these results support the once controversial hypothesis

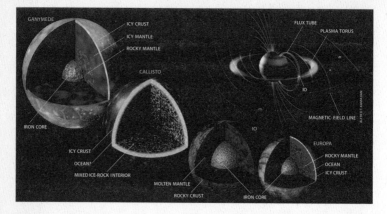

Jupiter's family of major moons each has a distinct personality and a very different interior. Of the four shown (not to scale), Io is by far the most interactive with Jupiter. Ionized gas from Io's volcanic eruptions are swept up by the planet's electrical field (inset), creating a powerful electrical current between the moon and the planet.

that the deep interior of Jupiter is deluged with helium rain. There helium becomes immiscible in the hydrogen-rich atmosphere, which at high pressures—millions of times sea-level pressure on Earth—is perhaps better thought of as an ocean. Being heavier, the helium gradually settles toward the center of the planet. Under certain conditions, neon dissolves in the helium raindrops. Helium may also precipitate out on Saturn, whose helium depletion may be even more extreme.

After several years of analysis, researchers recently announced the abundance of the other noble gases. Argon, krypton and xenon are enriched compared with the solar composition by about the same factor as carbon and sulfur. That, too, is a mystery. The only way to trap the inferred quantities of these gases is to freeze them—which is not possible at Jupiter's current distance from the sun. Therefore, much of the material that makes up the planet must have come from colder, more distant regions. Jupiter itself may even have formed farther from the sun, then drifted inward.

A final clue to Jovian history came from the measurement of deuterium, one of the heavy isotopes of hydrogen. The concentration is similar to that on the sun and is distinctly different from that of comets or of Earth's oceans. The finding suggests that comets have not had a major effect on the composition of Jupiter's atmosphere, despite the spectacular effects when they hit, as demonstrated during the Shoemaker-Levy 9 collisions in 1994.

After the orbiter relayed the probe data to Earth, it commenced its tour of the Jovian system—to date, a total of 26 orbits of the planet, with multiple flybys of each of the four Galilean satellites. The limelight has been on Europa, whose surface geology and other features point to the existence of a liquid ocean beneath the ice sometime in Europa's history, probably in the geologically recent past. But the other moons have not been neglected.

The innermost Galilean satellite, Io, stole the show during the two *Voyager* encounters. The initial pictures from those spacecraft showed a remarkably young surface, the only one in the solar system with essentially no impact craters. Later, images taken for navigation purposes serendipitously caught immense eruptive plumes. Subsequent observations confirmed that Io is wracked by volcanic activity. The size of Earth's moon, it spews 100 times more lava than Earth does.

Galileo has spent less time looking at Io than at the other moons, primarily because of the danger to the spacecraft: Io lies deep in Jupiter's intense radiation belts. *Galileo* flew within 900 kilometers of Io's surface just before the orbit insertion in 1995 but did not revisit until last October, when the bulk of its mission had been completed and scientists felt free to take more risks. Although concerns about the jam-prone tape recorder forced cancellation of imaging and spectroscopy during the 1995 flyby, the particle detector and magnetometer remained active.

They found that the empty space around Io is anything but. It seethes with subatomic particles blasted out by volcanic eruptions and stirred up by Jupiter's magnetic field. Electron beams course down the field lines that connect Io to Jupiter's atmosphere; dense, cold plasmas permeate the wake left behind Io by the magnetic field sweeping by. Whenever Io passed through Jupiter's shadow, *Galileo* saw the moon outlined by a thin ring of glowing gas, lit up by the impact of electrons from the Jovian magnetosphere. In short, Io is tightly linked to the giant planet by what amounts to the largest electric circuit in the solar system.

For most of its mission *Galileo* studied the tortured surface of Io from a safe distance. Based on how brightly the volcanoes glow at different visible and near-infrared wavelengths, it inferred their temperature, a measurement critical to determining the composition of the lavas. Most volcanoes on Earth disgorge lava of basaltic composition—iron, magnesium and calcium silicates rich in the minerals olivine and pyroxene. Basaltic melts typically have temperatures ranging from 1,300 to 1,450 kelvins (1,050 to 1,200 degrees Celsius). In contrast, telescopic observations of Io several years ago suggested temperatures of 1,500 to 1,800 kelvins. These temperatures ruled out substances that melt at lower temperatures, such as liquid sulfur, which had been suggested previously as a dominant volcanic fluid on Io.

When *Galileo's* measurements came down, the enigma intensified. Lavas on the moon are actually 1,700 to 2,000 kelvins. Magma this hot has not been common on Earth for more than three billion years. Io may thus be giving scientists an unexpected glimpse into Earth's geologic youth, a time when its interior temperatures were higher and the composition of the upper mantle different from today's.

When *Galileo* finally returned to Io last fall, the mission team was uncertain whether the spacecraft would survive the radiation. On one of its passes, it autonomously aborted the data-taking sequence just four hours before reaching Io, and the team rebooted with only minutes to spare. Several instruments also suffered damage, but all continued to work and in the end returned spectacular data. Io's active volcanoes were finally captured up close and personal.

One of *Galileo's* major discoveries was made during its very first orbital encounter—with Ganymede, Jupiter's largest moon. About half an hour before the spacecraft reached its closest approach, the radio-noise instrument, designed to record ambient electrical fields, began to go haywire. The relatively quiet background radio signals seen throughout most of the Jovian system changed abruptly to a complex, active radio spectrum. For 45 minutes the activity remained intense, and then it ceased as suddenly as it had begun. When the radio noise commenced, the magnetometer readings shot up fivefold.

Plasma researchers had seen signatures of this sort before, when spacecraft carrying similar instruments entered and exited magnetospheres at Earth, Jupiter, Saturn, Uranus and Neptune. Two subsequent Ganymede flybys confirmed their suspicions: the moon is magnetized, generating a dipole field similar to those of these planets. No other satellite has such a field. Earth's moon and Mars may have had fields in the past, but currently they exhibit only limited patches of magnetic variation that represent magnetized rocks on the surface. Like a set of nested Russian dolls, Ganymede has a magnetosphere contained within Jupiter's huge magnetic domain, which in turn is embedded in the sun's.

Tracking of the spacecraft signal allowed researchers to probe Ganymede's gravity field and therefore its internal structure. They concluded that it probably has a dense core about 1,500 kilometers in radius with a surrounding icy mantle 700 kilometers deep. Geochemical models suggest that the core consists of a sphere of iron or iron sulfide enveloped in rock. The inner metallic core could produce the dipolar magnetic field.

Yet theorists are not sure quite how. Although scientists compare planetary magnetic fields to bar magnets, the analogy can be misleading. Solid iron at the center of a planet or large moon would be too hot to retain a permanent magnetic field. Instead a magnetic field is thought to involve a convecting, conductive liquid. Models of Ganymede indicate that its interior can easily become hot enough to melt iron or iron sulfide. But the same models show that convection will cease as the core gradually cools; the conditions required for convection should last only a billion years or so.

The answer may lie in the orbital resonance of the inner three Galilean satellites. Io goes around Jupiter precisely four times for each time Europa completes two circuits and Ganymede one. Like pushing a child's swing in time with its natural pendulum period, this congruence allows small forces to accumulate into large outcomes—in this case, distorting the orbits from their default circular shape into more oblong ellipses. The effect on the moons is profound. Because the distance between them and Jupiter is continuously changing, the influence of Jupiter's gravity waxes and wanes, stretching the moons by an ever varying amount. The process, known as tidal heating, drives the volcanism on Io and keeps Europa's putative ocean from freezing.

Researchers used to think that tidal heating was of little consequence for Ganymede, the outermost of these three moons. But now they realize that the orbits may have shifted over time. Consequently, the resonances may once have been stronger and Ganymede's orbit more perturbed than it is now. The immense fault systems that wind across the surface may record this earlier period of intense heating. If so, the moon is still cooling off, and its core can continue to generate a magnetic field.

Compared with flamboyant Europa, Io and Ganymede, the outermost Galilean satellite, Callisto, was always thought rather drab. In *Voyager* images it epitomized the traditional stereotype for icy satellites: an old, frozen, pockmarked mudball. But *Galileo* observations tell a different story.

Callisto is covered with large impact scars, ranging from craters kilometers in diameter to the so-called palimpsest named Valhalla, some 1,500 kilometers across. The surface is believed to date back more than four billion years to the rain of meteoritic and cometary debris left after the formation of the planets and satellites. In this sense, Callisto is indeed old. Seen close-up, however, Callisto's surface is blanketed by fine, dark debris. Small craters, which on most other bodies are produced in abundance, are largely absent. Surface features appear softened and eroded. Clearly, some young processes have been at work. Among the ideas proposed have been electrostatic levitation of fine dust, which would allow it to "flow" across the surface, and evaporation of ices from the surface, which would leave behind deposits of darker, less volatile material. So far none of the explanations is satisfying.

Intriguingly, near-infrared spectra show not only water ice and hydrated minerals, as expected, but also four unusual absorption features near a wavelength of four microns. One appears to be carbon dioxide trapped in the surface, perhaps as inclusions in icy particles or bubbles produced by radiation damage to the surface. Two other spectral features probably represent sulfur in the surface, which may originate in Io's volcanic eruptions. The fourth spectral feature is the strangest. Its wavelength corresponds to that absorbed by carbon-nitrogen bonds. In fact, laboratory spectra of complex organic molecules called tholins by the late Carl Sagan are similar. Tholins are thought to resemble organic material in the solar nebula; clouds of interstellar ice grains have comparable spectra. Taken together, the data provide the first direct evidence that icy satellites contain the carbon, nitrogen and sulfur compounds common in primitive meteorites and comets. These materials are also some of the most important for life.

The internal structure of Callisto shows the same paradoxical dichotomy between age and youth that the surface exhibits. Unlike the other Galilean satellites, Callisto seems more like a uniformly dense sphere, indicating that most of its rock and ice are mixed together. A core is ruled out. Therefore, the interior has never been heated strongly, either by radioactive decay or by tides. The moon does not participate in the orbital resonance that kneads the other Galilean satellites.

On the other hand, the moon is far from dead. As the *Galileo* magnetometer found, Callisto seems to perturb the surrounding Jovian magnetic field in a peculiar pattern. This disturbance, unlike Ganymede's, resembles what is seen in classic physics experiments in which a hollow copper sphere is subjected to a changing magnetic field. In such an experiment, electric currents are set up in the conducting shell of the sphere, which in turn produces a magnetic field that exactly counters the imposed field. Callisto's field seems to be induced in much the same way.

But what could form the electrically conducting layer? Rock, ice and ionospheric particles are poor conductors. Researchers are left with a possibility that not long ago seemed outrageous: salty ocean water. Seawater is a weak conductor with the right properties to explain the readings. A global liquid layer some tens of kilometers thick could produce the observed signature. The combination of evidence for a comparatively undifferentiated interior and for a global ocean presents a severe challenge for theorists. Somehow Callisto must be hot enough to support an ocean but not so hot that light and heavy materials separate. The water layer might be sandwiched between a radioactively heated interior, where convection keeps the material mixed, and a thin icy shell, where a different convection cycle cools the ocean. So much for dull old Callisto.

Although much of *Galileo's* mission involved studies of the Galilean moons, the orbiter did not overlook the smaller members of Jupiter's family. Its camera captured each of the four inner, small moons—Metis, Adrastea, Amalthea and Thebe, in order of distance from Jupiter. A major finding was that these small moons are directly responsible for Jupiter's rings. A special series of pictures was taken while the spacecraft was within Jupiter's shadow, allowing the sun to backlight the tiny dust particles that make up the rings. These pictures not only show the main rings and the tenuous gossamer ring seen by *Voyager* in 1979, but also reveal for the first time the complex structure of the gossamer ring. It consists of multiple layers directly related to the orbits of Amalthea and Thebe. Thus, the rings are probably microscopic debris kicked off the moons by the impact of tiny meteoroids.

The data gathered by *Galileo* have revolutionized scientists' view of Jupiter and its moons, which we have come to recognize as a kind of planetary system comparable in complexity to the solar system itself. The *Voyager* flybys provided the adrenaline rush of seeing worlds for the first time, but only an intensive investigation such as *Galileo's* could have revealed the nuances and the limitations of seemingly straightforward categories such as "thundercloud" and "icy satellite." Soon it will be Saturn's turn to enter this new phase of exploration. Another two-in-one spacecraft—*Cassini-Huygens*—arrives there in 2004. It, too, will probably raise more questions than it answers.

Saturn, Jupiter, Uranus and Neptune—the so-called "outer planets"— all sport rings. What forms these elegant structures and what keeps them in place? We now recognize that planetary rings, once thought to be static, are in fact dynamically evolving structures, and that no two planets share the same ring structure.

is from the perspective of
e blue planet that we view
of wonder of the night ski
dreds of thousands of point
rkling lights twinkle overh
the night sky. Some are fa
y stars, some are far away
es that are composed of mil
million of stars, and som
sun's reflection off the c

Bejeweled Worlds

Joseph A. Burns,
Douglas P. Hamilton and
Mark R. Showalter

uch of the modern world's economy is based on inventions made possible by 19th-century physicist James Clerk Maxwell, father of electromagnetism and pioneer of thermodynamics. In terms of raw economic benefit, though, not much can be said for another of Maxwell's favorite subjects: the rings of Saturn. Apart from inspiring the sales of executive desk toys, planetary rings do not contribute conspicuously to the material wealth of nations. And yet that does not blunt their appeal. In his 1857 Adams Prize essay, Maxwell wrote:

There are some questions in Astronomy to which we are attracted . . . on account of their peculiarity . . . [rather] than from any direct advantage which their solution would afford to mankind. . . . I am not aware that any practical use has been made of Saturn's Rings . . . [b]ut when we contemplate the Rings from a purely scientific point of view, they become the most remarkable bodies in the heavens, except, perhaps, those still less *useful* bodies—the spiral [galaxies]. . . . When we have actually seen that great arch swung over the equator of the planet without any visible connection, we cannot bring our minds to rest.

A century and a half later Saturn's rings remain a symbol of all that is exotic and wondrous about the universe. Better observations have only heightened their allure. The findings of the past two decades have so overturned previous knowledge that essentially a new ring system—one much more complex and interesting than theory, observation or imagination had suggested—has been revealed.

Other giant planets besides Saturn have rings, and no two systems look alike. Rings are strange, even by the standards of astronomy. They are sculpted by processes that can be feeble and counterintuitive. For example, in rings, gravity can effectively repel material. We now appreciate that rings, once thought to be static, are continually evolving. We have seen the vital symbiosis between satellites and rings. Most important, we have recognized that planetary rings are more than just exquisite phenomena. Like Maxwell, modern scientists see analogies between rings and galaxies; in a very fundamental way, rings may also afford a glimpse into the solar system's ancient beginnings.

Saturn's rings, initially spied in 1610 by Galileo Galilei and interpreted as a planet-encircling hoop five decades later by Christiaan Huygens, stood alone for more than three and a half centuries. Then, in a span of just seven years, rings were discovered around the other three giant planets. Uranus's were detected first, in 1977. James L. Elliot, then at Cornell University, monitoring a star's brightness as Uranus crossed in front of it, noticed the signal blinking on and off. He inferred that a series of narrow bands, slightly elliptical or inclined, circumscribe the planet. In 1979 the *Voyager 1* spacecraft sighted Jupiter's diaphanous rings. Finally, in 1984, a technique like

Elliot's detected pieces of rings—but not full rings—around Neptune.

Those heady days passed, and ring research stagnated until the mid-1990s. Since then, a new era of ring exploration has begun. Observations have poured in from the Hubble Space Telescope, ground-based telescopes and the *Galileo* probe in orbit about Jupiter. And in July 2004 the *Cassini* spacecraft will begin its four-year tour of the Saturnian system.

Although the four known ring systems differ in detail, they share many general attributes. They are all richly textured, made up of multiple concentric rings often separated by gaps of various widths. Each ring is composed of innumerable particles—chunks of rock and ice—that independently circle the central planet while gently jostling one another. Rings fall into two general categories based on how densely packed the particles are, as described by the optical depth, a measure of the exponential decay of light as it penetrates perpendicularly through the ring. For the densest rings, such as Saturn's main rings (designated A and B) and the Uranian rings (designated by numbers and Greek letters), the optical depth can be as high as 4, which means that a mere 2 percent of the light leaks through. The most tightly packed of these rings contain particles that range from a few centimeters to several meters in diameter.

Particles in a dense ring system collide frequently, often several times during each orbit around the planet. In the process, energy is lost and angular momentum is redistributed. Because particles nearer to the planet move at a higher speed than do particles farther out, collisions hold back the inner particles (which then fall toward the planet) and push forward the outer ones (which then move away from the planet). Thus, a ring tends to spread radially. But the spreading takes time, and in this regard, a ring may be thought of as a viscous fluid that slowly diffuses inward and outward. Saturn's rings have an effective kinematic viscosity like that of air.

The energy loss, combined with angular-momentum redistribution, causes a dense ring system to flatten. Whatever its initial shape, the system quickly becomes a thin, near-equatorial disk. Saturn's rings are only tens of meters from top to bottom even though they stretch across several hundred thousand kilometers; they are proportionally as thick as a sheet of tissue paper spread over a football field.

Another consequence of dense packing is to strengthen the particles' own mutual gravitational attraction. This may be why Uranus's rings are slightly out of round: their self-gravity resists the tendency to smear into a circular band.

At the other extreme, the faintest known rings, such as Jupiter's rings and Saturn's outermost rings, have optical depths between 10^{-8} and 10^{-6}. Particles are as spread out as baseball outfielders. Because they collide infrequently, they tend not to settle into a flat disk. As we know from how these rings scatter light, the particles are fine dust, typically microns in size, comparable to the size of smoke particles. So these structures are literally smoke rings. The particles display unusual dynamics because, being so small, they are significantly affected by electromagnetic and radiation forces in addition to gravity.

Neptune's rings do not fall into this neat dichotomy; their optical depth lies between the two extremes. The Neptunian system is anomalous in other respects as well. Its densest ring is not a smooth band; it contains discontinuous arcs that together encompass less than a tenth of the circumference. Without some confinement mechanism at work, these structures should spread fully around the planet in about a year. Yet recent observations find that the positions of the arcs have shifted little in the past 15 years.

All dense ring systems nestle close to their planets, extending no farther than the so-called Roche limit, the radius within which the planet's tidal forces overwhelm the tendency of ring particles to agglomerate into larger bodies. Just outside the Roche limit is a zone where small, irregularly shaped moons can coexist with the rings. The interactions between rings and ring moons are implicated in many of the strangest aspects of rings.

For example, Saturn's E ring reaches across a broad region that encompasses the satellites Mimas, Tethys, Dione and Rhea, peaking in brightness at the orbit of the smooth, icy moon Enceladus. The narrow F ring, a tangle of several lumpy strands, sits isolated just beyond Saturn's A ring and also is straddled by two moons, Pandora and Prometheus. Correlations of satellite positions and ring features occur in the Jovian, Uranian and Neptunian systems as well.

Explaining how satellites wield such power has been the major advance in ring science over the past two decades. Three basic processes appear to be at work. The first is the orbital resonance, a tendency of gravitational forces to be magnified at positions where a particle's orbital period matches ratio of a satellite's orbital period.

Orbits that lie near resonant locations suffer unusually large distortions because the gentle tugs of moons are repeated systematically and therefore build up over time. Resonances are stronger for particles in orbits near a moon, but when the orbits are too close, different resonances vie for control, and motions become chaotic.

Throughout Saturn's enormous rings, only a few dozen ring locations respond to strong satellite resonances.

The outcome of these resonant perturbations varies. Strong ones clear material, accounting for the outer edges of Saturn's A and B rings. In some places, gaps are opened. Such a resonance may account for Neptune's discontinuous ring. Analogous resonances explain the distribution of material in the asteroid belt, for which the sun plays the role of the planet and Jupiter plays the role of the satellite.

Elsewhere in the A ring, resonances generate waves. If the satellite has an elliptical orbit, the result is a spiral wave, a miniature version of the pinwheel pattern of our galaxy. If the satellite has a tilted orbit, the result is a series of vertical bending waves, an out-of-plane corrugation—small ripples in a cosmic carpet.

Although resonances typically involve satellites, any force that repeats periodically at an integer ratio of the orbital period—such as lumpy planetary gravitational fields or variable electromagnetic forces—will be similarly effective. The Jovian system has become infamous for such resonances. Inward of a radius of 120,000 kilometers, the ring abruptly puffs up from a flat disk to a thick torus. A ring particle at that radius orbits three times for every two planetary spins; thus, the planet's tilted magnetic field pushes it ever upward. Still closer to the planet, at a radius of 100,000 kilometers, the brightness of the Jovian ring drops sharply. That happens to be the location of the 2:1 electromagnetic resonance. Particles that drift to this position are spread so thinly that they vanish against the giant planet's glare.

The second basic way that satellites govern ring structures is by influencing the paths of ring particles. The gravitational interaction of a satellite and a nearby particle is somewhat counterintuitive. If these two bodies were isolated in deep space, their close encounters would be symmetrical in space and time. The particle would approach the satellite, accelerate, zip around, emerge on the other side and decelerate (assuming it did not collide). The departure leg would be the mirror image of the inbound path (a hyperbola or parabola). Although the particle would have changed direction, it would eventually return to its original speed.

In a ring system, however, a satellite and particle are not isolated—they are in orbit around a third object, the planet. Whichever body is nearer to the planet orbits faster. Suppose it is the particle. During the close encounter, the gravity of the satellite nudges the particle into a new orbit. The event is asymmetrical: the particle moves closer to the satellite, and the gravitational interaction of the two bodies strengthens. So the particle is unable to regain the velocity it once had; its orbital energy and angular momentum have decreased. Technically, that means its orbit is distorted from a circle to an ellipse of slightly smaller size; later, collisions within the ring will restore the orbit to a circle, albeit a shrunken one.

The net effect is that the particle is pushed inward. Its loss is the satellite's gain, although because the satellite is more massive, it moves proportionately less. If the positions are reversed, so are the roles: with the satellite on the inside, the particle will be pushed outward and the satellite inward. In both cases, the attractive gravity of a satellite appears to *repulse* ring material. None of Newton's laws have been broken; this bizarre outcome occurs when two bodies in orbit around a third interact and lose energy.

Like resonances, this mechanism can pry open gaps in rings. The gaps will grow until the satellite's repulsive forces are counterbalanced by the tendency of rings to spread during collisions. Such gaps are present within Saturn's A, C and D rings, as well as throughout the Cassini division, a zone that separates the A and B rings.

Conversely, the process can squeeze a narrow ring. Satellites on either side of a strand of material can shepherd that material, pushing back any particles that try to escape. In 1978 Peter Goldreich and Scott D. Tremaine, then both at the California Institute of Technology, hypothesized the shepherding process to explain the otherwise puzzling stability of the threadlike rings of Uranus. The satellites Cordelia and Ophelia keep Uranus's ε ring corralled. Saturn's F ring appears to be herded by Prometheus and Pandora. To be sure, most of the visible gaps and narrow ringlets remain unexplained. Perhaps they are manipulated by moons too small to see with present technology. The *Cassini Orbiter* may be able to spy some of the hidden puppeteers.

Yet another effect of repulsive gravity is to scallop ring edges. These undulations are easiest to understand from the vantage point of the satellite. In rings, a continuous stream of particles flows past the satellite. When these particles overtake the moon, gravity modifies their circular orbits into elliptical ones of almost the same size. The particles no longer maintain a constant distance from the planet. Someone riding on the satellite would say that the particles have started to weave back and forth in concert. The apparent motion is sinusoidal with a wavelength proportional to the distance between the orbits of the satellite and the particle.

The resulting wave appears behind the satellite if the particle is on the outside and in front of the satellite if the particle is on the inside. It is akin to the wake of a boat in an unusual river where the water on one side of the boat moves faster than the boat itself. One of us (Showalter) analyzed the scalloped edges of Saturn's Encke division to pinpoint a small satellite, Pan, that had eluded observers. Another example is the F ring, whose periodic clumps seem to have been imprinted by Prometheus.

The third and final effect of moons on rings is to spew out and soak up material. This role, especially vital for faint, dusty rings such as those around Jupiter, has come into clear view only with the *Galileo* mission to Jupiter. Earlier the *Voyager* spacecraft had discovered Jupiter's rings as well as two small moons, Adrastea and Metis, close to the main ring's outer edge. But its camera was not sharp enough to tell us what the satellites actually did. Were they shepherds that prevented the rings' outward spread? Or were they the source of ring material that, once placed into orbit, drifted inward? Neither could *Voyager* make sense of a faint outer extension—a gossamer ring that accompanied the main one.

Galileo's imaging system found that the gossamer ring vanished abruptly beyond the orbit of the moon Amalthea. It discovered another, fainter gossamer ring that extended as far as the moon Thebe and no farther. On the flight home from the meeting at which these images were first available, one of us (Burns) noticed the smoking gun: the vertical extent of the innermost gossamer ring was equal to the orbital tilt of Amalthea, and the thickness of the outer gossamer ring perfectly matched the inclination of Thebe. Furthermore, both gossamer rings were brightest along their top and bottom edges, indicating a pileup of material—which is exactly what one would expect if particles and satellites shared the same orbital tilt. This tight association is most naturally explained if the particles are debris ejected by meteoroid impacts onto the satellites.

Ironically, small moons should be better sources of material than big ones: though smaller targets, they have weaker gravity, which lets more debris escape. In the Jovian system the most effective supplier is calculated to be 10 or 20 kilometers across—just about the size of Adrastea and Metis, explaining why they generate more formidable rings than do Amalthea and Thebe, which are much larger.

An odd counterexample is Saturn's 500-kilometer-wide moon, Enceladus, which appears to be the source of the E ring. Powerful impacts by ring particles, as opposed to interplanetary projectiles, might explain how Enceladus manages to be so prolific. Each grain that hits Enceladus generates multiple replacement particles, so the E ring could be self-sustaining. Elsewhere such collisions usually result in a net absorption of material from the ring.

The evident importance of sources and sinks reopens the classic question of whether rings are old and permanent or young and fleeting. The former possibility implies that rings could date to the formation of the solar system. Just as the protosun was surrounded by a flattened cloud of gas and dust out of which the planets are thought to have emerged, each of the giant planets was surrounded by its own cloud, out of which satellites emerged. Close to each planet, within the Roche limit, tidal forces prevented material from agglomerating into satellites. That material became a ring instead.

Alternatively, the rings we see today may have arisen much later. A body that strayed too close to a planet may have been torn asunder, or a satellite may have been shattered by a high-speed comet. Once a satellite is blasted apart, the fragments will reagglomerate only if they lie beyond the Roche limit. Even then, they will be unconsolidated, weak rubble piles susceptible to later disruption.

Several lines of evidence now suggest that most rings are young. First, tiny grains must lead short lives. Even if they survive interplanetary micrometeoroids and fierce magnetospheric plasma, the subtle force exerted by radiation causes their orbits to spiral inward. Unless replenished, faint rings should disappear within just a few thousand years. Second, some ring moons lie very close to the rings, even though the back reaction from spiral density waves should quickly drive them off.

Third, icy ring particles should be darkened by cometary debris, yet they are generally bright. Fourth, satellites just beyond Saturn's rings have remarkably low densities, as though they are rubble piles. Finally, some moons are embedded within rings. If rings are simply primordial material that failed to agglomerate, how did those moons get there? The moons make most sense if they are merely the largest remaining pieces of a shattered progenitor.

So it seems that rings are not quite the timeless fixtures they appear to be. Luke Dones of the Southwest Research Institute in Boulder, Colo., has suggested that Saturn's elaborate adornments are the debris of a shattered moon roughly 300 to 400 kilometers across. Whether all rings have such a violent provenance, we now know they were not simply formed and left for us to admire. They continually reinvent themselves. Joshua E. Colwell and Larry W. Esposito of the University of Colorado envision recycling of material between rings and ring moons. Satellites gradually sweep up the particles and subsequently slough them off during energetic collisions. Such an equilibrium could determine the extent of many rings. Variations in the composition, history and size of the planets and satellites would naturally account for the remarkable diversity of rings.

Indeed, the emerging synthesis explains why most of the inner planets are ringless: they lack large retinues of satellites to provide ring material. Earth's moon is too big, and any micronsize dust that does escape its surface is usually stripped away by solar gravitational and radiation forces. Mars, with its two tiny satellites, probably does have rings. But two of us (Hamilton and Showalter) were unable to find any rings or smaller satellites in Hubble observations in 2001. If a Martian ring does exist, it must be exceedingly tenuous, with an optical depth of less than 10^{-8}.

As often happens in science, the same basic principles apply to phenomena that at first seem utterly unrelated. The solar system and other planetary systems can be viewed as giant, star-encircling rings. Astronomers have seen hints of gaps and resonances in the dusty disks around other stars, as well as signs that source bodies orbit within. The close elliptical orbits of many large extrasolar planets are best understood as the end result of angular momentum transfer between these bodies and massive disks. Planetary rings are not only striking, exquisite structures; they may be the Rosetta stones to deciphering how planets are born.

Over the past ten years, the planet Pluto has gotten very little respect from Earth-bound observers. It was suggested that Pluto may not even qualify as a planet at all. Then it and its moon Charon, very close in size, have been dubbed by some as a "twin-planet" configuration. Most recently, Pluto has been called a mere object in the Kuiper belt. But a new space probe, slated for a 2015 arrival, may clear up the confusion.

Journey to the Farthest Planet

S. Alan Stern

Until about 10 years ago, most planetary scientists considered Pluto to be merely an oddity. All the other planets neatly fit into what astronomers knew about the architecture of the solar system—four small, rocky bodies in the inner orbits and four gas giants in the outer orbits, with an asteroid belt in between. But distant Pluto was an icy enigma traveling in a peculiar orbit beyond Neptune. Some researchers, most notably Dutch-American astronomer Gerard Kuiper, had suggested in the 1940s and 1950s that perhaps Pluto was not a world without context but the brightest of a vast ensemble of objects orbiting in the same region. This concept, which came to be known as the Kuiper belt, rattled around in the scientific literature for decades. But repeated searches for this myriad population of frosty worlds came up empty-handed.

In the late 1980s, however, scientists determined that something like the Kuiper belt was needed to explain why many short-period comets orbit so close to the plane of the solar system. This circumstantial evidence for a distant belt of bodies orbiting in the same region as Pluto drove observers back to their telescopes in search of faint, undiscovered objects beyond Neptune. By the 1980s telescopes were being equipped with electronic light detectors that made searches far more sensitive than work done previously with photographic plates. As a result, success would come their way.

In 1992 astronomers at the Mauna Kea Observatory in Hawaii discovered the first Kuiper belt object (KBO), which was found to be about 10 times as small as and almost 10,000 times as faint as Pluto [see "The Kuiper Belt," page 111]. Since then, observers have found more than 600 KBOs, with diameters ranging from 50 to almost 1,200 kilometers. (Pluto's diameter is about 2,400 kilometers.)

And that's just the tip of the iceberg, so to speak. Extrapolating from the small fraction of the sky that has been surveyed so far, investigators estimate that the Kuiper belt contains approximately 100,000 objects larger than 100 kilometers across. As a result, the Kuiper belt has turned out to be the big brother to the asteroid belt, with far more mass, far more objects (especially of large sizes), and a greater supply of ancient, icy and organic material left over from the birth of the solar system.

It is now clear that Pluto is not an anomaly. Instead it lies within a vast swarm of smaller bodies orbiting between about five billion and at least eight billion kilometers from the sun. Because this far-off region may hold important clues to the early development of the solar system, astronomers are keenly interested in learning more about Pluto, its moon, Charon, and the bodies making up the Kuiper belt. Unfortunately, the immense distance between this region of the solar system and Earth has limited the quality of observations. Even the exquisite Hubble Space Telescope, for example, shows only blurry regions of light and dark on Pluto's surface. And although the *Pioneer*, *Voyager* and *Galileo* spacecraft have provided scientists with marvelous close-up images of Jupiter, Saturn, Uranus

- Astronomers have recently learned that Pluto is not an anomaly, as once believed, but the brightest of a vast ensemble of objects orbiting in a distant region called the Kuiper belt. Scientists want to explore Pluto and the Kuiper belt objects because they may hold clues to the early history of the planets.
- Pluto and its moon, Charon, are also intriguing in their own right. The two bodies are so close in size that astronomers consider them a double planet. In addition Pluto has a rapidly escaping atmosphere and complex seasonal patterns.
- NASA has chosen a team called New Horizons to build a spacecraft that would study Pluto, Charon and several Kuiper belt objects during a series of flyby encounters. If its funding is approved by Congress, the spacecraft could be launched in 2006 and arrive at Pluto as early as 2015.

and Neptune, no space probe has ever visited the Pluto-Charon system or the Kuiper belt.

Recognizing the importance of this region of the solar system, scientists have urged NASA to put Pluto on its planetary exploration agenda for more than a decade. In response, the space agency has studied mission concepts ranging from houseboat-size, instrument-laden spacecraft similar to the *Cassini* probe (now on its way to Saturn) to hamster-size craft carrying just a camera. In the late 1990s NASA settled on a midsize concept called *Pluto-Kuiper Express* that would be built by the Jet Propulsion Laboratory in Pasadena, Calif. But the projected cost of that mission quickly rose toward $800 million, which was considerably more than NASA wanted to invest. As a result, the agency scrapped *Pluto-Kuiper Express* in the fall of 2000.

But this cancellation didn't go down easily. Scientists, space exploration advocates and schoolchildren flooded NASA with requests to reconsider, and the agency did so, but with a twist. Rather than restarting the expensive *Pluto-Kuiper Express*, NASA launched a competition among universities, research labs and aerospace companies for proposals to explore Pluto, Charon and the Kuiper belt at lower cost. Never before had NASA allowed industry and universities to compete to lead a mission to the outer solar system. Given the novelty of such a competition, NASA made it clear that if none of the proposals could accomplish the specified scientific measurement objectives by 2020, and for less than $500 million, then the agency was under no obligation to choose *any* of them.

After a grueling selection process, NASA picked our team, called New Horizons, to carry out the Pluto–Kuiper belt mission. New Horizons is led by my institution, the Southwest Research Institute, based in San Antonio, Tex., and the Applied Physics Laboratory (APL) at *Johns Hopkins* University. A team of scientists from more than a dozen universities, research institutions and NASA centers is deeply involved in planning the scientific observations. The Southwest Research Institute will manage the project, be in charge of the mission team and be responsible for the development of the scientific instruments. APL will build and operate the *New Horizons* spacecraft. Ball Aerospace, the NASA Goddard Space Flight Center and Stanford University will build portions of the instrument payload, and JPL will be responsible for spacecraft tracking and navigation.

By pioneering less expensive ways to build and operate a spacecraft to explore the outer solar system, *New Horizons* satisfied NASA's conditions: the total mission cost is $488 million, including more than $80 million in budgeted reserves, and the spacecraft may arrive at Pluto as early as the summer

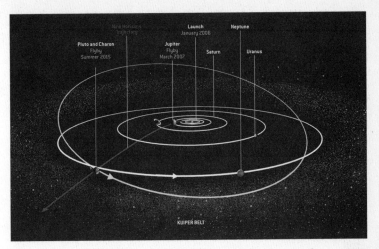

KUIPER BELT

Exploration of the Kuiper belt could take place as early as 2006 if the planned *New Horizons* spacecraft is launched on schedule. After a flyby of Jupiter, a gravity-assisted slingshot effect would carry the craft to Pluto by 2015, and later out into the Kuiper belt to investigate this distant region of icy bodies and debris.

of 2015. Furthermore, *New Horizons* would fly more instruments and return about 10 times as much observational data as the canceled *Pluto-Kuiper Express* mission would have delivered and would do so for less money.

Why are astronomers so interested in studying Pluto-Charon and the Kuiper belt? I can summarize only a few of the reasons here. For one, the size, shape, mass and general nature of the Kuiper belt appear to be much like the debris belts seen around other nearby stars, such as Vega and Fomalhaut. Researchers, including myself, have used computer modeling techniques to simulate the formation of the KBOs almost five billion years ago as the planetary system was coalescing from a whirling disk of gas and dust. We found that the ancient Kuiper belt must have been approximately 100 times as massive as it is today to give rise to Pluto-Charon and the KBOs we see. In other words, there was once enough solid material to have formed another planet the size of Uranus or Neptune in the Kuiper belt.

The same simulations also revealed that large planets like Neptune would have naturally grown from the KBOs in a very short time had nothing disturbed the region. Clearly, something disrupted the Kuiper belt at about the time Pluto was formed, but we do not yet know the cause of the disturbance. Perhaps it was the formation of Neptune near the belt's inner boundary. Did the planet's gravitational influence somehow interrupt the creation of another gas giant farther out? And if so, why didn't the formation of Uranus frustrate the birth of Neptune in the same way? Perhaps instead it was the gravitational influence of a large population of planetary embryos—rocky bodies thousands of kilometers across—moving rapidly through the Kuiper belt billions of years ago after they were ejected by Uranus and Neptune from their formation zones. Or perhaps it was something else altogether. Whatever the cause, the Kuiper belt lost most of its mass, and the growth of bodies in the region was suddenly arrested.

The KBOs are remnants of that ancient planet-building process and therefore hold extremely important clues to the formation of the outer solar system. Exploring Pluto and the Kuiper belt is the equivalent of conducting an archaeological dig into the history of the outer solar system—a place where researchers can get a valuable glimpse of the long-gone era of planetary formation.

Furthermore, although our knowledge of Pluto and Charon is meager, what we do know indicates that they offer a scientific wonderland of their own. For one, Charon is surprisingly large—its diameter is about 1,200 kilometers, or about half of Pluto's. Because the two bodies are so close in size, Pluto-Charon can be considered a double planet. No other planet in our solar system falls into this category—the diameters of most satellites are just a few percent of the diameters of their parent planets. But because astronomers have discovered many double asteroids and double KBOs in recent years, there is now little doubt that binary objects like Pluto-Charon are common in our solar system and most likely in others. Yet we have never visited a binary world.

We are eager to know how a system such as Pluto-Charon could form. The prevailing theory is that Pluto collided with another large body in the distant past and that much of the debris from this impact went into orbit around Pluto and eventually coalesced to form Charon. Because it appears that a similar collision led to the creation of Earth's moon, the study of Pluto and Charon is expected to shed some light on that subject as well.

Researchers also want to know why Pluto and Charon are so different in appearance. Observations from Earth and the Hubble Space Telescope show that Pluto has a highly reflective surface with distinct markings that indicate the presence of expansive polar caps. In contrast, Charon's surface is far less reflective, with indistinct markings. And whereas Pluto has an atmosphere, Charon apparently does not. Is the sharp dichotomy between these two neighboring worlds a result of divergent evolution, perhaps because of their different sizes and compositions, or is it a consequence of how they originally formed? We do not know.

Also intriguing is the fact that Pluto's density, size and surface composition are strikingly similar to those of Neptune's largest satellite, Triton. One of the great surprises of *Voyager 2's* exploration of the Neptune system was the discovery of ongoing, vigorous volcanic activity on Triton. Will Pluto also display such activity? Will the KBOs as well? Our present-day knowledge of planetary processes suggests that they should not, but Triton's activity was not expected either. Perhaps Triton is showing us that we do not yet understand the nature of small worlds. By exploring Pluto and the KBOs, we expect to gain a better comprehension of this fascinating class of bodies.

Yet another of Pluto's alluring features is its bizarre atmosphere. Although Pluto's atmosphere is about 30,000 times less dense than Earth's, it offers some unique insights into the workings of planetary atmospheres. Whereas Earth's atmosphere contains only one gas (water vapor) that regularly undergoes phase transitions between solid and gaseous states, Pluto's atmosphere contains three: nitrogen, carbon monoxide and methane. Furthermore, the current temperature on Pluto varies by about 50 percent across its surface—from about 40 to about 60 kelvins. Pluto reached its closest approach to the sun in 1989. As the planet moves farther away, most astronomers believe that the average surface temperature will drop and that most of the atmosphere will condense and fall as snow. Pluto may well have the most dramatic seasonal patterns of any planet in the solar system.

What is more, Pluto's atmosphere bleeds into space at a rate much like a comet's. Most of the molecules in the upper atmosphere have enough thermal energy to escape the planet's gravity; this extremely fast leakage is called hydrodynamic escape. Although this phenomenon is not seen on any other planet today, it may have been responsible for the rapid loss of hydrogen from Earth's atmosphere early in our planet's history. In this way, hydrodynamic escape may have helped make Earth suitable for life. Pluto is now the only planet in the solar system where this process can be studied.

An important connection between Pluto and the origin of life on Earth is the presence of organic compounds, such as frozen methane, on Pluto's surface and water ice in its interior. Recent observations of KBOs show that they, too, probably harbor large quantities of ice and organics. Billions of years ago such objects are believed to have routinely strayed into the inner part of the solar system and helped to seed the young Earth with the raw materials of life.

* * *

Given so many compelling scientific motivations, it is not hard to understand why the planetary research community wants to send a spacecraft to Pluto and the Kuiper belt. And given the romance and adventure of exploring uncharted worlds, it is not surprising that so many citizens and grade school children have also become excited about this mission to new frontiers.

NASA's request for Pluto–Kuiper belt mission proposals specified three top priorities for scientific observations. First, the craft must map the surfaces of Pluto and Charon with an average resolution of one kilometer (in contrast, the Hubble Space Telescope cannot do better than about 500-kilometer resolution when it views Pluto and Charon). Second, the probe must map the surface composition across the various geologic provinces of the two bodies. Third, the craft must determine the composition and structure of Pluto's atmosphere, as well as its escape rate. NASA also outlined a list of lower priorities, including the measurement of surface temperatures and the search for additional satellites or rings around Pluto. The agency also required that the spacecraft accomplish the same objectives for at least one KBO beyond Pluto.

When NASA selected our proposal late in 2001, it stated that the *New Horizons* mission offered both the best scientific return and the lowest risk of schedule delays and cost over-runs. This was, in part, because of the robust capabilities of the spacecraft we proposed and the experience of our team-member institutions at delivering space missions on schedule and at or below cost.

The *New Horizons* spacecraft we designed is lean, with a planned mass of just 416 kilograms (917 pounds)—heavier than the early *Pioneer* probes but lighter than the *Voyagers*. This mass includes the hydrazine maneuvering propellant that will be used to adjust the craft's trajectory in flight. Most of the spacecraft's subsystems, such as its computers and its propulsion-control system, are based on designs used in the

APL's *Comet Nucleus Tour* (CONTOUR) probe, which is scheduled to launch this July on a multiple comet flyby mission. The use of CONTOUR's designs reduces *New Horizons's* costs and lowers the risk of both technical and schedule problems. Almost all our spacecraft subsystems include spare equipment to increase reliability on the long flight to Pluto and the Kuiper belt.

The spacecraft will carry four instrument packages. A mapping and compositional spectroscopy package, PERSI, will make observations in the visible, ultraviolet and infrared parts of the spectrum. PERSI's infrared imaging spectrometer will be essential for mapping the composition and physical state (including temperature) of the surface ices on Pluto and Charon. A radio-science instrument dubbed REX will probe Pluto's atmospheric structure and gauge the average surface temperatures of Pluto and Charon (on both the daysides and nightsides of the bodies) by measuring the intensity of the microwave radiation striking the spacecraft's 2.5-meter-wide radio dish. A third instrument suite, PAM, consists of charged-particle detectors designed to sample material escaping from Pluto's atmosphere and to determine its escape rate. The fourth instrument is LORRI, a high-resolution imager that will supplement PERSI's already formidable mapping capabilities. At closest approach, PERSI's global maps of Pluto-Charon and the KBOs will have an average resolution of one kilometer. But LORRI, which will image selected regions, will be able to detect objects 20 times as small!

If all goes as planned, the spacecraft will be launched in January 2006, heading initially for a flyby of Jupiter that will use the planet's gravity to slingshot the craft toward Pluto. During its Jupiter flyby, *New Horizons* will conduct an intensive four-month study of the planet's intriguing system of more than 20 moons, as well as its auroras, atmosphere and magnetosphere. Thanks to the gravitational assist from Jupiter, the spacecraft can reach the Pluto-Charon system as early as 2015.

For much of the long cruise from Jupiter to Pluto, *New Horizons* will slumber in electronic hibernation. Turning off unneeded systems and reducing the amount of contact with the craft lowers the chance of equipment failures and drastically decreases mission operations costs. During this hibernation the craft will continuously transmit a simple status beacon to Earth; if an unexpected problem develops, our ground-control team will respond. Once each year the craft will be awakened for about 50 days to thoroughly test the systems, make course corrections and calibrate its scientific instruments.

Unlike earlier plans for a quick flyby of Pluto-Charon, *New Horizons* will begin its study of Pluto-Charon six months before its closest approach to the planet. Once the craft is about 100 million kilometers from Pluto—about 75 days before closest approach—its images of the planet will be better than those of the Hubble Space Telescope, and the results will improve with each passing day. In the weeks before closest approach, our mission team will be able to map Pluto-Charon in increasing detail and observe phenomena such as Pluto's weather by comparing the images of the planet over time. And using LORRI's high-resolution imaging capabilities, we will get "zoom-lens" views of Pluto and Charon that will help us decide which geologic features are worthy of special scrutiny. During the day of closest approach, when *New Horizons* may come as near as a few thousand kilometers from Pluto, PERSI will obtain its best maps of the entire sunlit faces of Pluto and Charon. Meanwhile LORRI will focus on producing higher-resolution maps of dozens of smaller areas on these bodies.

Once the spacecraft passes Pluto, it will turn around and map the planet's nightside, which will be softly illuminated by the reflected moonlight from Charon. And the spacecraft's antenna will receive a powerful radio beam from Earth passing through Pluto's atmosphere. By measuring the refraction of this radio beam, we will be able to plot the temperature and density profile of Pluto's atmosphere from high altitude down to the surface.

After the Pluto-Charon encounter, *New Horizons* will almost immediately maneuver to begin a series of what we hope will be three or more similar flybys with ancient KBOs over the next five years. The exact number of encounters will depend on how much propellant is left in the spacecraft after the Pluto flyby.

The *New Horizons* mission promises to revolutionize our knowledge of both the Pluto-Charon system and the Kuiper belt. But the potential for discovery will be lost if the mission is not launched in 2006. Because of the changing alignment of the planets, after 2006 the spacecraft will no longer be able to accelerate toward Pluto by swinging past Jupiter. If this window is missed, NASA would have to wait until 2018 for Jupiter to be in the right place again, delaying any encounter until the mid-2020s at the earliest.

By that time Pluto will be hundreds of millions of kilometers farther from the sun and significantly colder than it is today. Because of a combination of Pluto's extreme polar tilt and its motion around the sun, more than four million kilometers of terrain—much of the planet's southern hemisphere—will by then be covered in a dark polar shadow, thereby preventing it from being observed. Also, it is likely that virtually all the planet's atmosphere will have condensed by then, closing off any opportunity to study it until the 23rd century, when the atmosphere should again rise as the planet makes its next close approach to the sun.

New Horizons represents a thrilling return to first-time exploration for NASA's planetary program: for the first time since 1989, when *Voyager* 2 flew by Neptune, a spacecraft will train its instruments on a new world. The mission offers a scientific bonanza reminiscent of NASA's historic explorations.

Long believed to occupy the fringes of the solar system outside the orbit of Neptune, the existence of the Kuiper belt is now an established scientific fact. Most members of the Kuiper belt are seen to be random rocky bodies along with some comet-like material, all probably left over from the birth of our solar system. Like the planets, objects in the belt orbit the sun in regular orbits and, when planetary alignments assert a gravitational pull on them, some may become short-term comets.

The Kuiper Belt

Jane X. Luu and
David C. Jewitt

After the discovery of Pluto in 1930, many astronomers became intrigued by the possibility of finding a 10th planet circling the sun. Cloaked by the vast distances of interplanetary space, the mysterious "Planet X" might have remained hidden from even the best telescopic sight, or so these scientists reasoned. Yet decades passed without detection, and most researchers began to accept that the solar system was restricted to the familiar set of nine planets.

But many scientists began seriously rethinking their notions of the solar system in 1992, when we identified a small celestial body—just a few hundred kilometers across—sited farther from the sun than any of the known planets. Since that time, we have identified nearly three dozen such objects circling through the outer solar system. A host of similar objects is likely to be traveling with them, making up the so-called Kuiper belt, a region named for Dutch-American astronomer Gerard P. Kuiper, who, in 1951, championed the idea that the solar sytem contains this distant family.

What led Kuiper, nearly half a century ago, to believe the

disk of the solar sytem was populated with numerous bodies orbiting at great distances from the sun? His conviction grew from a fundamental knowledge of the behavior of certain comets—masses of ice and rock that on a regular schedule plunge from the outer reaches of the solar system inward toward the sun. Many of these comparatively small objects periodically provide spectacular appearances when the sun's rays warm them enough to drive dust and gas off their surfaces into luminous halos (creating large "comae") and elongate tails.

Astronomers have long realized that such active comets must be relatively new members of the inner solar system. A body such as Halley's comet, which swings into view every 76 years, loses about one ten-thousandth of its mass on each visit near the sun. That comet will survive for only about 10,000 orbits, lasting perhaps half a million years in all. Such comets were created during the formation of the solar system 4.5 billion years ago and should have completely lost their volatile constituents by now, leaving behind either inactive, rocky nuclei or diffuse streams of dust. Why then are so many comets still around to dazzle onlookers with their displays?

The comets that are currently active formed in the earliest days of the solar system, but they have since been stored in an inactive state—most of them preserved within a celestial deep freeze called the Oort cloud. The Dutch astronomer Jan H. Oort proposed the existence of this sphere of cometary material in 1950. He believed that this cloud had a diameter of about 100,000 astronomical units (AU—a distance defined as the average separation between Earth and the sun, about 150 million kilometers) and that it contained several hundred billion individual comets. In Oort's conception, the random gravitational jostling of stars passing nearby knocks some of the outer comets in the cloud from their stable orbits and gradually deflects their paths to dip toward the sun.

For most of the past half a century, Oort's hypothesis neatly explained the size and orientation of the trajectories that the so-called long-period comets (those that take more than 200 years to circle the sun) follow. Astronomers find that those bodies fall into the planetary region from random directions—as would be expected for comets originating in a spherical repository like the Oort cloud. In contrast, Oort's hypothesis could not explain short-period comets that normally occupy smaller orbits tilted only slightly from the orbital plane of Earth—a plane that astronomers call the ecliptic.

Most astronomers believed that the short-period comets originally traveled in immense, randomly oriented orbits (as the long-period comets do today) but that they were diverted by the gravity of the planets—primarily Jupiter—into their current orbital configuration. Yet not all scientists subscribed to this idea. As early as 1949, Kenneth Essex Edgeworth, an Irish gentleman-scientist (who was not affiliated with any research institution) wrote a scholarly article suggesting that there could be a flat ring of comets in the outer solar system. In his 1951 paper, Kuiper also discussed such a belt of comets, but he did not refer to Edgeworth's previous work.

Kuiper and others reasoned that the disk of the solar system should not end abruptly at Neptune or Pluto (which vie with each other for the distinction of being the planet most distant from the sun). He envisioned instead a belt beyond Neptune and Pluto consisting of residual material left over from the formation of the planets. The density of matter in this outer region would be so low that large planets could not have accreted there, but smaller objects, perhaps of asteroidal dimensions, might exist. Because these scattered remnants of primordial material were so far from the sun, they would maintain low surface temperatures. It thus seemed likely that these distant objects would be composed of water and ice and various frozen gases—making them quite similar (if not identical) to the nuclei of comets.

Kuiper's hypothesis languished until the 1970s, when Paul

C. Joss of the Massachusetts Institute of Technology began to question whether Jupiter's gravity could in fact efficiently transform long-period comets into short-period ones. He noted that the probability of gravitational capture was so small that the large number of short-period comets that now exists simply did not make sense. Other researchers were, however, unable to confirm this result, and the Oort cloud remained the accepted source of the comets, long and short period alike.

But Joss had sown a seed of doubt, and eventually other astronomers started to question the accepted view. In 1980 Julio A. Fernández (then at the Max Planck Institute for Aeronomy in Katlenburg-Lindau) had, for example, done calculations that suggested that short-period comets could come from Kuiper's proposed trans-Neptunian source. In 1988 Martin J. Duncan of the University of Toronto, Thomas Quinn and Scott D. Tremaine (both at the Canadian Institute for Theoretical Astrophysics) used computer simulations to investigate how the giant gaseous planets could capture comets. Like Joss, they found that the process worked rather poorly, raising doubts about the veracity of this well-established concept for the origin of short-period comets. Indeed, their studies sounded a new alarm because they noted that the few comets that could be drawn from the Oort cloud by the gravitational tug of the major planets should be traveling in a spherical swarm, whereas the orbits of the short-period comets tend to lie in planes close to the ecliptic.

Duncan, Quinn and Tremaine reasoned that short-period comets must have been captured from original orbits that were canted only slightly from the ecliptic, perhaps from a flattened belt of comets in the outer solar system. But their so-called Kuiper belt hypothesis was not beyond question. In order to make their calculations tractable, they had exaggerated the masses of the outer planets as much as 40 times (thereby increasing the amount of gravitational attraction and speeding up the orbital evolution they desired to examine). Other astrophysicists wondered whether this computational sleight of hand might have led to an incorrect conclusion.

Even before Duncan, Quinn and Tremaine published their work, we wondered whether the outer solar system was truly empty or instead full of small, unseen bodies. In 1987 we began a telescopic survey intended to address exactly that question. Our plan was to look for any objects that might be present in the outer solar system using the meager amount of sunlight that would be reflected back from such great distances. Although our initial efforts employed photographic plates, we soon decided that a more promising approach was to use an electronic detector (a charge-coupled device, or CCD) attached to one of the larger telescopes.

We conducted the bulk of our survey using the University of Hawaii's 2.2-meter telescope on Mauna Kea. Our strategy was to use a CCD array with this instrument to take four sequential, 15-minute exposures of a particular segment of the sky. We then enlisted a computer to display the images in the sequence in quick succession—a process astronomers call "blinking." An object that shifts slightly in the image against the background of stars (which appear fixed) will reveal itself as a member of the solar system.

For five years, we continued the search with only negative results. But the technology available to us was improving so rapidly that it was easy to maintain enthusiasm (if not funds) in the continuing hunt for our elusive quarry. On August 30, 1992, we were taking the third of a four-exposure sequence while blinking the first two images on a computer. We noticed that the position of one faint "star" appeared to move slightly between the successive frames. We both fell silent. The motion was quite subtle, but it seemed definite. When we compared the first two images with the third, we realized that we had indeed found something out of the ordinary. Its slow motion across the sky indicated that the newly discovered object could be traveling beyond even the outer reaches of Pluto's distant orbit. Still, we were suspicious that the mysterious object might be a near-Earth asteroid moving in parallel with Earth (which might also cause a slow apparent motion). But further measurements ruled out that possibility.

We observed the curious body again on the next two nights and obtained accurate measurements of its position, brightness and color. We then communicated these data to Brian G. Marsden, director of the International Astronomical Union's Central Bureau of Astronomical Telegrams at the Smithsonian Astrophysical Observatory in Cambridge, Mass. His calculations indicated that the object we had discovered was indeed orbiting the sun at a vast distance (40 AU)—only slightly less remote than we had first supposed. He assigned the newly discovered body a formal, if somewhat drab, name based on the date of discovery: he christened it "1992 QB_1." (We preferred to call it "Smiley," after John Le Carré's fictional spy, but that name did not take hold within the conservative astronomical community.)

Our observations showed that QB_1 reflects light that is quite rich in red hues compared with the sunlight that illuminates it. This odd coloring matched only one other object in the solar system—a peculiar asteroid or comet called 5145 Pholus. Planetary astronomers attribute the red color of 5145 Pholus to the presence of dark, carbon-rich material on its surface. The similarity between QB_1 and 5145 Pholus thus heightened our excitement during the first days after the discovery. Perhaps the object we had just located was coated by some kind of red material abundant in organic compounds. How big was this ruddy new world? From our first series of measurements, we estimated that QB_1 was between 200 and 250 kilometers across—about 15 times the size of the nucleus of Halley's comet.

Some astronomers initially doubted whether our discovery of QB_1 truly signified the existence of a population of objects in the outer solar system, as Kuiper and others had hypothesized.

But such questioning began to fade when we found a second body in March 1993. This object is as far from the sun as QB_1 but is located on the opposite side of the solar system. During the past three years, several other research groups have joined the effort, and a steady stream of discoveries has ensued. The current count of trans-Neptunian, Kuiper belt objects is 32.

The known members of the Kuiper belt share a number of characteristics. They are, for example, all located beyond the orbit of Neptune, suggesting that the inner edge of the belt may be defined by this planet. All these newly found celestial bodies travel in orbits that are only slightly tilted from the ecliptic—an observation consistent with the existence of a flat belt of comets. Each of the Kuiper belt objects is millions of times fainter than can be seen with the naked eye. The 32 objects range in diameter from 100 to 400 kilometers, making them considerably smaller than both Pluto (which is about 2,300 kilometers wide) and its satellite, Charon (which measures about 1,100 kilometers across).

The current sampling is still quite modest, but the number of new solar system bodies found so far is sufficient to establish beyond doubt the existence of the Kuiper belt. It is also clear that the belt's total population must be substantial. We estimate that the Kuiper belt contains at least 35,000 objects larger than 100 kilometers in diameter. Hence, the Kuiper belt probably has a total mass that is hundreds of times larger than the well-known asteroid belt between the orbits of Mars and Jupiter.

The Kuiper belt may be rich in material, but can it in fact serve as the supply source for the rapidly consumed short-period comets? Matthew J. Holman and Jack L. Wisdom, both then at M.I.T., addressed this problem using computer simulations. They showed that within a span of 100,000 years the gravitational influence of the giant gaseous planets (Jupiter, Saturn, Uranus and Neptune) ejects comets orbiting in their vicinity, sending them out to the farthest reaches of the solar system. But a substantial percentage of trans-Neptunian comets can escape this fate and remain in the belt even after 4.5 billion years. Hence, Kuiper belt objects located more than 40 AU from the sun are likely to have held in stable orbits since the formation of the solar system.

Astronomers also believe there has been sufficient mass in the Kuiper belt to supply all the short-period comets that have ever been formed. So the Kuiper belt seems to be a good candidate for a cometary storehouse. And the mechanics of the transfer out of storage is now well understood. Computer simulations have shown that Neptune's gravity slowly erodes the inner edge of the Kuiper belt (the region within 40 AU of the sun), launching objects from that zone into the inner solar system. Ultimately, many of these small bodies slowly burn up as comets. Some—such as Comet Shoemaker-Levy 9, which collided with Jupiter in July 1994—may end their lives suddenly by striking a planet (or perhaps the sun). Others will be caught in a gravitational slingshot that ejects them into the far reaches of interstellar space.

If the Kuiper belt is the source of short-period comets, another obvious question emerges: Are any comets now on their way from the Kuiper belt into the inner solar system? The answer may lie in the Centaurs, a group of objects that includes the extremely red 5145 Pholus. Centaurs travel in huge planet-crossing orbits that are fundamentally unstable. They can remain among the giant planets for only a few million years before gravitational interactions either send them out of the solar system or transfer them into tighter orbits.

With orbital lifetimes that are far shorter than the age of the solar system, the Centaurs could not have formed where they currently are found. Yet the nature of their orbits makes it practically impossible to deduce their place of origin with certainty. Nevertheless, the nearest (and most likely) reservoir is the Kuiper belt. The Centaurs may thus be "transition comets," former Kuiper belt objects heading toward short but showy lives within the inner solar system. The strongest evidence supporting this hypothesis comes from one particular Centaur—2060 Chiron. Although its discoverers first thought it was just an unusual asteroid, 2060 Chiron is now firmly established as an active comet with a weak but persistent coma.

As astronomers continue to study the Kuiper belt, some have started to wonder whether this reservoir might have yielded more than just comets. Is it coincidence that Pluto, its satellite, Charon, and the Neptunian satellite Triton lie in the vicinity of the Kuiper belt? This question stems from the realization that Pluto, Charon and Triton share similarities in their own basic properties but differ drastically from their neighbors.

The densities of both Pluto and Triton, for instance, are much higher than any of the giant gaseous planets of the outer solar system. The orbital motions of these bodies are also quite strange. Triton revolves around Neptune in the "retrograde" direction—opposite to the orbital direction of all planets and most satellites. Pluto's orbit slants highly from the ecliptic, and it is so far from circular that it actually crosses the orbit of Neptune. Pluto is, however, protected from possible collision with the larger planet by a special orbital relationship known as a 3:2 mean-motion resonance. Simply put, for every three orbits of Neptune around the sun, Pluto completes two.

The pieces of the celestial puzzle may fit together if one postulates that Pluto, Charon and Triton are the last survivors of a once much larger set of similarly sized objects. S. Alan Stern of the Southwest Research Institute in Boulder first suggested this idea in 1991. These three bodies may have been swept up by Neptune, which captured Triton and locked Pluto—perhaps with Charon in tow—into its present orbital resonance.

Interestingly, orbital resonances appear to influence the position of many Kuiper belt objects as well. Up to one half of the newly discovered bodies have the same 3:2 mean-motion resonance as Pluto and, like that planet, may orbit serenely for billions of years. (The resonance prevents Neptune from approaching too closely and disturbing the orbit of the smaller body.) We have dubbed such Kuiper belt objects Plutinos—"little Plutos." Judging from the small part of the sky we have examined, we estimate that there must be several thousand Plutinos larger than 100 kilometers across.

The recent discoveries of objects in the Kuiper belt provide a new perspective on the outer solar system. Pluto now appears special only because it is larger than any other member of the Kuiper belt. One might even question whether Pluto deserves the status of a full-fledged planet. Strangely, a line of research that began with attempts to find a 10th planet may, in a sense, have succeeded in reducing the final count to eight. This irony, along with the many intriguing observations we have made of Kuiper belt objects, reminds us that our solar system contains countless surprises.

Beyond even the reaches of the Kuiper belt lies the Oort cloud, the "Siberia" of our solar system. Named for Dutch astronomer Jan H. Oort who theorized its existence in 1950, the Oort cloud has yet to be seen by Earth-bound scientists but its existence is inferred by the presence of long-term comets. Most researchers believe this far-off region of the solar system is composed mostly of objects made of frozen gas and ice.

The Oort Cloud

Paul R. Weissman

I t is common to think of the solar system as ending at the orbit of the most distant known planet, Pluto. But the sun's gravitational influence extends more than 3,000 times farther, halfway to the nearest stars. And that space is not empty— it is filled with a giant reservoir of comets, leftover material from the formation of the solar system. That reservoir is called the Oort cloud.

The Oort cloud is the Siberia of the solar system, a vast, cold frontier filled with exiles of the sun's inner empire and only barely under the sway of the central authority. Typical noon-time temperatures are a frigid four degrees Celsius above absolute zero, and neighboring comets are typically tens of millions of kilometers apart. The sun, while still the brightest star in the sky, is only about as bright as Venus in the evening sky on Earth.

We have never actually "seen" the Oort cloud. But no one has ever seen an electron, either. We infer the existence and properties of the Oort cloud and the electron from the physical effects we can observe. In the case of the former, those effects are the steady trickle of long-period comets into the planetary system. The existence of the Oort cloud answers questions that people have asked since antiquity: What are comets, and where do they come from?

Aristotle speculated in the fourth century B.C. that comets were clouds of luminous gas high in Earth's atmosphere. But the Roman philosopher Seneca suggested in the first century A.D. that they were heavenly bodies, traveling along their own paths through the firmament. Fifteen centuries passed before his hypothesis was confirmed by Danish astronomer Tycho Brahe, who compared observations of the comet of 1577 made from several different locations in Europe. If the comet had been close by, then from each location it would have had a slightly different position against the stars. Brahe could not detect any differences and concluded that the comet was farther away than the moon.

Just how much farther started to become clear only when astronomers began determining the comets' orbits. In 1705 the English astronomer Edmond Halley compiled the first catalogue of 24 comets. The observations were fairly crude, and Halley could fit only rough parabolas to each comet's path. Nevertheless, he argued that the orbits might be very long ellipses around the sun:

> For so their Number will be determinate and, perhaps, not so very great. Besides, the Space between the Sun and the fix'd Stars is so immense that there is Room enough for a Comet to revolve, tho' the Period of its Revolution be vastly long.

In a sense, Halley's description of comets circulating in orbits stretching between the stars anticipated the discovery of the Oort cloud two and a half centuries later. Halley also noticed that the comets of 1531, 1607 and 1682 had very similar orbits and were spaced at roughly 76-year intervals. These seemingly different comets, he suggested, were actually the same comet returning at regular intervals. That body, now known as Halley's comet, last visited the region of the inner planets in 1986.

Since Halley's time, astronomers have divided comets into two groups according to the time it takes them to orbit the sun (which is directly related to the comets' average distance from the sun). Long-period comets, such as the recent bright comets Hyakutake and Hale-Bopp, have orbital periods greater than 200 years; short-period comets, less than 200 years. In the past decade astronomers have further divided the short-period comets into two groups: Jupiter-family comets, such as comets Encke and Tempel 2, which have periods less than 20 years; and intermediate-period, or Halley-type, comets, with periods between 20 and 200 years.

These definitions are somewhat arbitrary but reflect real differences. The intermediate- and long-period comets enter the planetary region randomly from all directions, whereas the Jupiter-family comets have orbits whose planes are typically inclined no more than 40 degrees from the ecliptic plane, the plane of Earth's orbit. (The orbits of the other planets are also very close to the ecliptic plane.) The intermediate- and long-period comets appear to come from the Oort cloud, whereas the Jupiter-family comets are now thought to originate in the Kuiper belt.

By the early 20th century, enough long-period cometary orbits were available to study their statistical distribution. A problem emerged. About one third of all the "osculating" orbits—that is, the orbits the comets were following at the point of their closest approach to the sun—were hyperbolic. Hyperbolic orbits would originate in and return to interstellar space, as opposed to elliptical orbits, which are bound by gravity to the sun. The hyperbolic orbits led some astronomers to suggest that comets were captured from interstellar space by encounters with the planets.

To examine this hypothesis, celestial-mechanics researchers extrapolated, or "integrated," the orbits of the long-period comets backward in time. They found that because of distant gravitational tugs from the planets, the osculating orbits did not represent the comets' original orbits. When the effects of the planets were accounted for—by integrating far enough back in time and orienting the orbits not in relation to the sun but in relation to the center of mass of the solar system (the sum of the sun and all the planets)—almost all the orbits became elliptical. Thus, the comets were members of the solar system, rather than interstellar vagabonds.

In addition, although two thirds of these orbits still appeared to be uniformly distributed, fully one third had orbital energies that fell within a narrow spike. That spike represented orbits that extend to very large distances—20,000 astronomical units (20,000 times the distance of Earth from the sun) or more. Such orbits have periods exceeding one million years.

Why were so many comets coming from so far away? In the late 1940s Dutch astronomer Adrianus F. van Woerkom showed that the uniform distribution could be explained by planetary perturbations, which scatter comets randomly to both larger and smaller orbits. But what about the spike of comets with million-year periods?

In 1950 Dutch astronomer Jan H. Oort, already famous for having determined the rotation of the Milky Way galaxy in the 1920s, became interested in the problem. He recognized that the million-year spike must represent the source of the long-period comets: a vast spherical cloud surrounding the planetary system and extending halfway to the nearest stars.

Oort showed that the comets in this cloud are so weakly bound to the sun that random passing stars can readily change their orbits. About a dozen stars pass within one parsec

(206,000 astronomical units) of the sun every one million years. These close encounters are enough to stir the cometary orbits, randomizing their inclinations and sending a steady trickle of comets into the inner solar system on very long elliptical orbits. As they enter the planetary system for the first time, the comets are scattered by the planets, gaining or losing orbital energy. Some escape the solar system altogether. The remainder return and are observed again as members of the uniform distribution. Oort described the cloud as "a garden, gently raked by stellar perturbations."

A few comets still appeared to come from interstellar space. But this was probably an incorrect impression given by small errors in the determination of their orbits. Moreover, comets can shift their orbits because jets of gas and dust from their icy surfaces act like small rocket engines as the comets approach the sun. Such nongravitational forces can make the orbits appear hyperbolic when they are actually elliptical.

Oort's accomplishment in correctly interpreting the orbital distribution of the long-period comets is even more impressive when one considers that he had only 19 well-measured orbits to work with. Today astronomers have more than 15 times as many. They now know that long-period comets entering the planetary region for the first time come from an average distance of 44,000 astronomical units. Such orbits have periods of 3.3 million years.

Astronomers have also realized that stellar perturbations are not always gentle. Occasionally a star comes so close to the sun that it passes right through the Oort cloud, violently disrupting the cometary orbits along its path. Comets close to the star's path are thrown out to interstellar space, while the orbits of comets throughout the cloud undergo substantial changes.

Although close stellar encounters have no direct effect on the planets—the closest expected approach of any star over the history of the solar system is 900 astronomical units from the sun—they might have devastating indirect consequences. In 1981 Jack G. Hills, now at Los Alamos National Laboratory, suggested that a close stellar passage could send a "shower" of comets toward the planets, raising the rate of cometary impacts on the planets and possibly even causing a biological mass extinction on Earth. According to computer simulations I performed in 1985 with Piet Hut, then at the Institute for Advanced Study in Princeton, N.J., the frequency of comet passages during a shower could reach 300 times the normal rate. The shower would last two to three million years.

Recently Kenneth A. Farley and his colleagues at the California Institute of Technology found evidence for just such a comet shower. Using the rare helium 3 isotope as a marker for extraterrestrial material, they plotted the accumulation of interplanetary dust particles in ocean sediments over time. The rate of dust accumulation is thought to reflect the number of comets passing through the planetary region; each comet sheds dust along its path. Farley discovered that this rate increased sharply at the end of the Eocene epoch, about 36 million years ago, and decreased slowly over two to three million years, just as theoretical models of comet showers would predict. The late Eocene is identified with a moderate biological extinction event, and several impact craters have been dated to this time. Geologists have also found other traces of impacts in terrestrial sediments, such as iridium layers and microtektites.

Is Earth in danger of a comet shower now? Fortunately not. Joan Garcia-Sanchez of the University of Barcelona, Robert A. Preston and Dayton L. Jones of the Jet Propulsion Laboratory in Pasadena, Calif., and I have been using the positions and velocities of stars, measured by the *Hipparcos* satellite, to reconstruct the trajectories of stars near the solar system. We have found evidence that a star has passed close to the sun in the past one million years. The next close passage of a star will occur in 1.4 million years, and that is a small red dwarf called

Gliese 710, which will pass through the outer Oort cloud about 70,000 astronomical units from the sun. At that distance, Gliese 710 might increase the frequency of comet passages through the inner solar system by .50 percent—a sprinkle perhaps, but certainly no shower.

In addition to random passing stars, the Oort cloud is now known to be disturbed by two other effects. First, the cloud is sufficiently large that it feels tidal forces generated by the disk of the Milky Way and, to a lesser extent, the galactic core. These tides arise because the sun and a comet in the cloud are at slightly different distances from the midplane of the disk or from the galactic center and thus feel a slightly different gravitational tug. The tides help to feed new long-period comets into the planetary region.

Second, giant molecular clouds in the galaxy can perturb the Oort cloud, as Ludwig Biermann of the Max Planck Institute for Physics and Astrophysics in Munich suggested in 1978. These massive clouds of cold hydrogen, the birthplaces of stars and planetary systems, are 100,000 to one million times as massive as the sun. When the solar system comes close to one, the gravitational perturbations rip comets from their orbits and fling them into interstellar space. These encounters, though violent, are infrequent—only once every 300 million to 500 million years. In 1985 Hut and Scott D. Tremaine, now at Princeton University, showed that over the history of the solar system, molecular clouds have had about the same cumulative effect as all passing stars.

Currently three main questions concern Oort-cloud researchers. First, what is the cloud's structure? In 1987 Tremaine, Martin J. Duncan, now at Queen's University in Ontario, and Thomas R. Quinn, now at the University of Washington, studied how stellar and molecular-cloud perturbations redistribute comets within the Oort cloud. Comets at its outer edge are rapidly lost, either to interstellar space or to the inner solar system, because of the perturbations. But deeper inside, there probably exists a relatively dense core that slowly replenishes the outer reaches.

Tremaine, Duncan and Quinn also showed that as comets fall in from the Oort cloud, their orbital inclinations tend not to change. This is a major reason why astronomers now think the Kuiper belt, rather than the Oort cloud, accounts for the low-inclination, Jupiter-family comets. Still, the Oort cloud is the most likely source of the higher-inclination, intermediate-period comets, such as Halley and Swift-Tuttle. They were probably once long-period comets that the planets pulled into shorter-period orbits.

The second main question is, How many comets inhabit the Oort cloud? The number depends on how fast comets leak from the cloud into interplanetary space. To account for the observed number of long-period comets, astronomers now estimate the cloud has six trillion comets, making Oort-cloud comets the most abundant substantial bodies in the solar system. Only a sixth of them are in the outer, dynamically active cloud first described by Oort; the remainder are in the relatively dense core. If the best estimate for the average mass of a comet—about 40 billion metric tons—is applied, the total mass of comets in the Oort cloud at present is about 40 times that of Earth.

Finally, from where did the Oort-cloud comets originally come? They could not have formed at their current position, because material at those distances is too sparse to coalesce. Nor could they have originated in interstellar space; capture of comets by the sun is very inefficient. The only place left is the planetary system. Oort speculated that the comets were created in the asteroid belt and ejected by the giant planets during the formation of the solar system. But comets are icy bodies, essentially big, dirty snowballs, and the asteroid belt was too warm for ices to condense.

A year after Oort's 1950 paper, astronomer Gerard P. Kuiper

of the University of Chicago proposed that comets coalesced farther from the sun, among the giant planets. (The Kuiper belt is named for him because he suggested that some comets also formed beyond the farthest planetary orbits.) Comets probably originated throughout the giant planets' region, but researchers used to argue that those near Jupiter and Saturn, the two most massive planets, would have been ejected to interstellar space rather than to the Oort cloud. Uranus and Neptune, with their lower masses, could not easily throw so many comets onto escape trajectories. But more recent dynamical studies have cast some doubt on this scenario. Jupiter and particularly Saturn do place a significant fraction of their comets into the Oort cloud. Although this fraction may be smaller than that of Uranus and Neptune, it may have been offset by the greater amount of material initially in the larger planets' zones.

Therefore, the Oort-cloud comets may have come from a wide range of solar distances and hence a wide range of formation temperatures. This fact may help explain some of the compositional diversity observed in comets. Indeed, recent work I have done with Harold F. Levison of the Southwest Research Institute in Boulder, Colo., has shown that the cloud may even contain asteroids from the inner planets' region. These objects, made of rock rather than ice, may constitute 2 to 3 percent of the total Oort-cloud population.

The key to these ideas is the presence of the giant planets, which hurl the comets outward and modify their orbits if they ever reenter the planetary region. If other stars have giant planets, as observations over the past few years suggest, they may have Oort clouds, too. If each star has its own cloud, then as stars pass by the sun, their Oort clouds will pass through our cloud. Even so, collisions between comets will be rare because the typical space between comets is an astronomical unit or more.

The Oort clouds around each star may slowly be leaking comets into interstellar space. These interstellar comets should be easily recognizable if they were to pass close to the sun, because they would approach the solar system at much higher velocities than the comets from our own Oort cloud. To date, no such interstellar comets have ever been detected. This fact is not surprising; because the solar system is a very small target in the vastness of interstellar space, there is at best a 50–50 chance that people should have seen one interstellar comet by now.

The Oort cloud continues to fascinate astronomers. Through the good fortunes of celestial mechanics, nature has preserved a sample of material from the formation of the solar system in this distant reservoir. By studying it and the cosmo-chemical record frozen in its icy members, researchers are learning valuable clues about the origin of the solar system.

Energy from the sun streams past the planets and extends far into space. The term heliosphere refers to that region in space where Sol's energy holds sway. But where are the limits to the heliosphere, and just how far-reaching is its influence? Four aging spacecraft are racing to the outer reaches of the solar system with the intent to travel past the sun's extended domain and into interstellar space.

Quest for the Limits of the Heliosphere

J. R. Jokipii and
Frank B. McDonald

Glowing comets and brilliant auroras are visible reminders that space in the inner solar system is far from empty. This region is permeated by swiftly flowing charged particles emanating from the sun, a continuous torrent of solar wind that often blows in sudden gusts. The fetch of this wind extends well past the orbit of the earth or the range of visible comets. The outward rush of particles and the solar magnetic field carried with them carve an enormous spherical cavity in the interstellar medium that reaches far beyond the orbit of the most distant planets of the solar system. This immense region, a bubble of solar dominance within the vastness of space, is called the heliosphere.

One might imagine that with increasing distance from the sun the heliosphere gradually fades to a diffuse boundary wherein particles of the solar wind gently mix with the interstellar breeze of dust and gas. But this is not at all the case: near the limit of the outer heliosphere lies an abrupt discontinuity at which a myriad of intriguing physical phenomena are thought likely to occur. As of yet, however, astrophysicists have no direct measurements of the heliosphere's outer margins and so must infer, theorize or simply speculate on its exact nature. We do not even know with any certainty how far from the sun this boundary forms. But our ignorance of the distant reaches of the heliosphere may last only a few more years, when space probes finally break through this first barrier toward interstellar space.

Past the orbits of Neptune and Pluto, on trajectories taking them beyond the edges of the solar system, drifts a small flotilla of spacecraft. This modest scientific armada consists of *Pioneer 10* and *11*, along with *Voyager 1* and *2*, all of which were launched about two decades ago. If we were to look back at the solar system from any of these spacecraft today, the sun would be the brightest object in view, but it would nonetheless appear more than 1,000 times dimmer than as seen from the earth. Even at these great distances, though, the four spacecraft remain well within the heliosphere. Onboard instruments continue to register disturbances originating on the sun's surface that propagate outward at about 400 kilometers per second. Despite this enormous velocity, these sudden gusts still take many months to reach the probes.

The original mission of the *Pioneer* and *Voyager* spacecraft—to study the giant planets Jupiter, Saturn, Uranus and Neptune—stands as one of the enduring triumphs of space exploration. But the continuing vitality of these four probes after the celebrated planetary flybys and our growing awareness of the complex and dynamic behavior of the distant solar wind have engendered an important second mission for these versatile scientific minions: to study the most remote parts of the heliosphere and its interface with the interstellar medium. The success of this newly established mission depends not only on the technical capabilities of the spacecraft and their earth-based controllers but also on the nature of the heliosphere itself.

The general structure of the solar wind and the heliosphere was first outlined three decades ago by Eugene N. Parker of

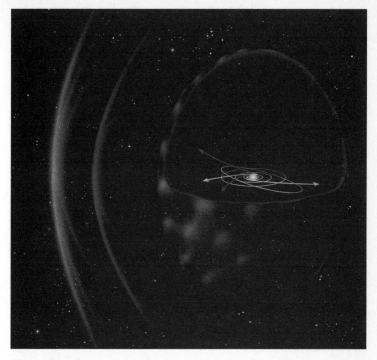

The heliosphere spans the area of space where the pressure of solar wind keeps intergallactic space at bay. Near the heliosphere's outer margins, a bow shock is believed to form. Data from the *Pioneer* and *Voyager* spacecraft should improve our understanding of the heliosphere's boundaries.

the University of Chicago in a series of revolutionary theoretical papers. Observations have since verified the structure he predicted, at least out to the position of the farthest spacecraft, *Pioneer 10*. We now have confirmed that the solar wind, as it moves radially outward from the sun, brings with it the imprint of the solar atmosphere: certain parts of the sun emit high-speed wind in vast streams that flow hundreds of kilometers per second faster than the typical wind. If the sun were stationary, those streams would simply form linear rays, but because it rotates every 27 days, fast streams originating near the solar equator can overtake slower streams emanating from adjacent areas on the surface. Sometimes this configuration lasts for several solar rotations, setting up regions in space where the interface between fast and slow streams of solar plasma becomes spiral in shape. These irregularities rotate with the sun and are known to space physicists as co-rotating interaction regions.

In addition, some parts of the solar atmosphere can eject irregular puffs and gusts that travel as transient disturbances in the wind. Occasionally, huge eruptions on the sun produce blast waves that severely disrupt the solar wind's more steady currents.

The many varied interactions of the solar-wind plasma produce shock waves, which heat the wind and also generate energetic particles. Co-rotating interaction regions and their associated shock waves are a major feature of the solar wind out to more than 10 astronomical units. (One astronomical unit, or AU, is the radius of the earth's orbit around the sun, some 150 million kilometers, or 93 million miles.) Farther out, such interaction regions combine, forming so-called global merged interaction regions, which populate space to the outer reaches of the heliosphere.

Embedded within the stream structure of the heliosphere lies a complex interplanetary magnetic field. The interplay of the magnetic field and solar wind can be rather complex; some of this behavior, however, can be readily visualized in terms of the familiar concept of magnetic lines of force and the properties these field lines give to the solar wind as it expands outward.

By earthly standards the plasma that constitutes the solar wind might seem rather insubstantial and formless. Yet because it is a good electrical conductor and because the kinetic energy of the flow is so much greater than the energy of the magnetic field, the magnetic-field lines in the heliosphere can be treated as though they move with the solar wind, being effectively "frozen in." This frozen magnetic flux lends the tenuous plasma added pressure and viscosity. These properties develop from magnetic forces rather than from the more familiar molecular interactions found in denser fluids. So the heliosphere contains a tangle of magnetic-field lines, stretched out by the wind into an enormous spiral whorl, which until recently had hardly been explored.

Early observations indicated that the magnetic-field lines alternated between being directed inward or outward from the sun at different solar longitudes. The first chance to measure the field at relatively high heliographic latitudes came in 1974, after *Pioneer 11* deflected its trajectory out of the ecliptic plane using Jupiter's large gravitational attraction to exchange momentum. It came as a surprise (although Michael Schulz, a space physicist at Aerospace Corporation, had predicted it) when *Pioneer 11* and *Voyager 1* reached the heliographic latitude of 16 degrees north and discovered that the magnetic field was nearly always directed outward. Edward J. Smith of the Jet Propulsion Laboratory in Pasadena, Calif., concluded that the spacecraft were observing magnetic fields carried by the solar wind and that at northern solar latitudes this field was oriented away from the sun.

The *Pioneer* and *Voyager* missions showed in 1976 that the sun's magnetic field was organized such that the field lines in the northern hemisphere generally pointed outward from the sun; those in the opposite hemisphere pointed inward. Because the polarity of the sun's field changes every 11 years (at the time of the sunspot maximum), a magnetic cycle lasting 22 years results. So in 1986 *Pioneer 11* and *Voyager 1* detected that the northern field was pointed duly inward.

In the heliosphere the transition between the inward- and outward-directed magnetic fields has the shape of a very thin, warped surface that is carried outward by the solar wind to form a vast interplanetary current sheet. Solar rotation twists the sheet so that the wrinkles lie along spiral magnetic-field lines and rotate with the sun. Space probes near the solar equatorial plane detect magnetic fields that are alternately directed inward and outward as the current sheet rotates past them in space. The sheet is least wrinkled during solar minima, the periods of fewer sunspots and lowered activity that occur every 11 years. Its geometry becomes so convoluted during solar maxima that the normal sheetlike structure becomes entirely unrecognizable.

But in whatever configuration, the magnetic-field pattern originating at the sun's surface is carried to the distant margins of the heliosphere by the solar wind over about a year's time. During this period, fast plasma streams continue to merge with slower ones, spawning regions of enhanced plasma density and magnetic-field strength. Out to some great but as yet unknown distance, the large-scale structure of the solar wind and the magnetic field is fundamentally preserved.

As it travels, the solar wind expands over an increasingly large volume. Eventually the solar-wind plasma is spread so thinly that it can no longer push outward against the small inward pressure of the local interstellar medium. The wind does not slow down gradually at this point, because its velocity is greater than that at which disturbances can move within it. Instead the solar wind undergoes a sudden, violent change in speed.

This behavior follows from the fundamentals of supersonic fluid flow. For the motion to diminish incrementally, the downstream material must signal the upstream fluid to slow. These signals must be carried by sound waves moving through the medium. But such waves cannot propagate against flows moving faster than sound. As a result, the upstream fluid crashes into the fluid ahead, setting up a confrontation called a shock wave. Something similar occurs in a highway accident when cars behind cannot slow down fast enough to avoid hitting those ahead.

Much as with a multicar accident, we expect the solar-wind termination shock to be irregular and turbulent. As the solar-wind gas passes through the shock, its outward velocity should slow to about one quarter of its original value. Some of the wind's kinetic energy is converted to heat, raising the temperature of the interstellar gas to more than a million degrees Celsius. Some kinetic energy goes into compressing the magnetic field: we expect that field strength should jump to about four times its value inside the shock. So at this boundary, where the solar wind trades outward velocity for heat and turbulence, we expect to find a giant, spheroidal shock front with a complex but still somewhat mysterious structure.

Before spacecraft offered direct measurements, astrophysicists relied mainly on the study of cosmic rays to deduce something of the nature of the outer heliosphere. Galactic cosmic rays are subatomic particles (electrons, protons, all the heavier nuclei from helium to uranium, positrons and a small number of antiprotons) that travel at close to the speed of light and appear to populate all parts of the universe. Their ubiquitous presence in the cosmos is inferred from their ability to produce high-energy gamma rays and radio waves (which can be detected on the earth). Within our galaxy, cosmic rays commonly originate where the shock-wave remnants of supernova explosions are thought to accelerate the particles to extremely high energies. In addition, during periods of heightened activity, the sun, too, can occasionally produce significant numbers of solar "cosmic" rays of lower energy.

The heliosphere is constantly bathed by galactic cosmic rays. These cosmic-ray particles can diffuse upwind against the solar plasma because of their extremely high speeds and the presence of irregularities in the field. Because of their electrical charge, the particles gyrate tightly around magnetic-field lines, and as a result, cosmic rays also tend to travel out of the heliosphere along with the frozen-in magnetic flux. In general,

Probes Built to Go the Distance

Deep-space missions present major technical challenges to spacecraft designers in the areas of weight, power and communications. The *Pioneer* missions called not only for escape from the earth's gravity but also for sufficient energy to reach Jupiter with a travel time of two years or less. The solution was to use a high-energy rocket booster and to keep the spacecraft as light as possible—at launch *Pioneer 10* weighed only 250 kilograms. Reliable sources of electricity were also critical: far from the sun, the only practical ones are radioisotope thermoelectric generators (RTGs), which use the decay of radioactive materials to produce electricity from heat. Each *Pioneer* spacecraft has four RTG units, which initially generated a total of 155 watts. The ultimate lifetime of these missions will probably be defined by the radioactive decay of the plutonium oxide fuel and the degradation of the conversion elements within the RTG units.

The *Voyager* probes were designed in the mid-1970s using what had been learned from the earlier *Pioneer* missions. A more powerful launch vehicle made it possible to deploy heavier (825 kilograms) and more complex spacecraft: the *Voyager* design includes significant onboard computer capability, an experiment platform with precision pointing and improved RTG units that supplied 470 watts at launch. The sophistication and flexibility of the *Voyager* system have been demonstrated by the considerable reengineering that was done in flight to prepare *Voyager 2* for its late-scheduled encounter with Uranus [see "Engineering *Voyager 2*'s Encounter with Uranus," by Richard P. Laeser, William I. McLaughlin and Donna M. Wolff; *Scientific American*, November 1986].

Communication with the *Pioneer* and *Voyager* probes demands a large onboard antenna, the dominant feature. The *Pioneer* craft support a 2.7-meter parabolic dish antenna, somewhat smaller than the 3.7-meter dish found on the *Voyager* probes. The *Pioneer* antenna spins about the spacecraft's axis, which is kept pointed toward the earth by the occasional use of small thrusters. The *Voyager* spacecraft do not spin but are stabilized about all three axes so that the high-gain antenna can be kept directed toward giant antennae on the earth.

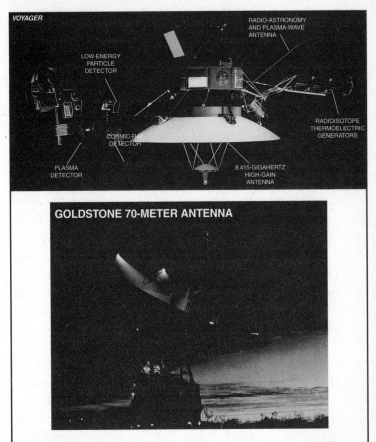

The three sites of NASA's Deep Space Network (in California, Australia and Spain) are among the most critical components of the *Pioneer* and *Voyager* programs. The Jet Propulsion Laboratory in Pasadena, Calif., has significantly upgraded this system by adding new receivers and increasing antenna size. For more reliable communication, the total rate of data transmission from *Pioneer 10* has been slowed to its minimum value of 16 bits a second—about as fast as one might send messages by Morse code. The *Voyagers'* larger antenna and higher transmission frequency make possible the use of the more available 34-meter antenna of the Deep Space Network and allow communication at 160 bits a second.

the solar wind acts to modulate the intensity of cosmic rays impinging on the earth, making it difficult for them to reach the inner heliosphere. This exclusion is most effective at low energies; highly energetic cosmic rays proceed largely unaffected. Because the fraction excluded varies with solar activity, cosmic-ray intensity follows the 11-year sunspot cycle: it peaks when solar activity is at a minimum.

Galactic cosmic rays pass through the outer heliosphere, and so they can provide valuable information about this unexplored region. Much of our understanding of the outer reaches of the solar system has been derived by comparing models of how cosmic rays traverse the heliosphere with observations. For example, data from the four deep-space probes have shown that the cosmic-ray gradient—the rate at which the intensity of galactic cosmic rays increases with heliocentric distance—is much smaller than was expected. This finding indicates that the heliosphere is larger than was predicted before the launch of the *Pioneer* and *Voyager* probes.

In the early 1970s, as the *Pioneer* spacecraft moved toward Jupiter, detectors on a number of spacecraft revealed the existence of an unexpected low-energy cosmic-ray component. Continuing study has demonstrated an enhancement at low energies for rays consisting of helium, nitrogen, oxygen, neon, argon and, most recently, hydrogen nuclei. This peculiar composition and energy spectrum define the anomalous cosmic-ray component. Observations from the *Pioneer* and *Voyager* spacecraft have shown that the intensity of the anomalous cosmic rays increases with distance from the sun.

What is the origin of this mysterious cosmic-ray component? Work over the past two decades has painted a compelling picture of how these cosmic rays are generated, although its accuracy is not completely proved.

In 1974 Lennard A. Fisk, Benzion Kozlovsky and Reuven Ramaty, while at the NASA Goddard Space Flight Center, suggested that the anomalous component originates as neutral atoms in interstellar space. As the heliosphere moves through the interstellar gas, neutral atoms, which are not affected by magnetic fields or other forces of the plasma, stream freely into the inner heliosphere. Those that pass near the sun are ionized by solar radiation or by the solar wind itself to become singly charged ions. Once the neutral atoms become ions, the magnetic-field lines in the solar wind snare them and convect these particles outward. Fisk and his colleagues speculated that subsequent acceleration to higher energies turns these ions into the anomalous cosmic rays.

The original basis for this suggestion was that most of the carbon in the interstellar medium cannot take part in this process, because carbon is almost completely ionized in interstellar space (which explains its very low abundance in the anomalous component). Recent observations near the earth by a number of space missions have demonstrated that the anomalous oxygen (and presumably also the other components) is singly charged. This result supports the model of Fisk and his co-workers: cosmic-ray nuclei from a nearby source (within the heliosphere) can retain some electrons, whereas normal cosmic rays are fully stripped of all their electrons during their passage through the galaxy.

But how were these newly formed ions accelerated to the observed cosmic-ray energies? During the 1970s, a number of proposals were put forth. None, however, successfully predicted the steady increase in the intensity of the anomalous cosmic rays registered by the *Pioneer* and *Voyager* probes as they moved far out into the heliosphere. Then, in 1981, one of us (Jokipii), along with Mark E. Pesses and David Eichler, both then at the University of Maryland, suggested that the acceleration of singly charged ions occurs at the termination-shock boundary. Plasma shocks can accelerate charged particles, and this location seemed a likely site for energizing the anomalous cosmic rays—it contains the strongest, most long-lived shock anywhere in the heliosphere. Detailed computer modeling has since shown that most observed features of the anomalous component follow naturally from this notion.

Important clues about the nature of the termination region have been collected by Donald A. Gurnett and William S. Kurth of the University of Iowa. Since 1983 they have registered low-frequency bursts of radio noise (at two to three kilohertz) using detectors on board both the *Voyager* spacecraft. The signals persist for many months and then gradually drift to higher frequencies. In July 1992 these researchers observed the onset of a particularly strong radio event and noted that it occurred more than 400 days after an unusually intense period of solar activity. This sequence followed the same pattern as another large noise burst in 1983. These remarkable radio signals probably originate just beyond the termination shock and, along with the anomalous cosmic rays, provide tantalizing information about this vast unexplored frontier.

As the *Pioneer* and *Voyager* space probes speed farther and farther from the sun, there is an increasing likelihood that they will soon encounter the termination shock. Estimates based on what was then known about the interstellar medium had originally put the termination-shock boundary anywhere from 75 to 150 AU from the sun, but data collected so far from the probes would suggest considerably smaller values. So it is entirely possible that one or more of the probes will reach the shock within the next decade.

NASA scientists have therefore taken steps to ensure that the proper measurements will be made during passage through the shock. Indeed, they believe they may have several opportunities to observe it, as gusts and turbulence in the solar wind move the termination shock in and out—perhaps leading to multiple crossings as the front moves back and forth past the spacecraft. Once the spacecraft finally pass beyond the shock, the wind will slacken, and, for the first time, an artifact of humanity will begin to experience directly the effects of the interstellar plasma. Perhaps then the true nature of the interstellar medium will finally be clarified.

Sometime in the 21st century, after having reported the physical conditions of the outer heliosphere and possibly the termination shock itself, the four spacecraft will continue their journey to the stars. *Pioneer 10* should remain operational until the turn of the century (at about 70 AU), and *Voyager 2* has enough consumables to last until about 2015 (at about 130 AU). But even after steerage and communication are lost, for eons to come these probes will follow a well-charted course through our galaxy as four small man-made objects added to the gaseous clouds of interstellar space. They go as the first voyagers from planet Earth, like small bottles tossed into an infinite sea.

is from the perspective of
e blue planet that we view
of wonder of the night ski
dreds of thousands of point
rkling lights twinkle overh
the night sky. Some are fa
y stars, some are far away
es that are composed of mil
millions of stars, and som
sun's reflection off the c

Conclusion

The sun rings like a bell. Acoustic currents churn its interior and power the creation of energy that floods the solar system. All planets, in some measure, receive light, heat, and energy from Sol. Its gravity marshals the planets into orderly orbits. Its solar wind partitions a tiny portion of the vast universe, separating Earth and her extended family of planets and satellites from the depths of interstellar space.

As we search our solar system and study our planetary neighbors both near and far, the uniqueness of our own planet is the first thing we see. On battered Mercury, broiling Venus, chilled Mars, and the odd outer planets composed primarily of gas, we find family members that are so dissimilar to our own planet that it makes us wonder: can there be life anywhere else in the universe?

The enormous diversity of planets in our solar system may be a reflection of the diversity and range of planets outside our solar system. It also challenges our definition of life. Are planets inert masses of thoughtless materials, or does the ceaseless geological activity we find on so many planets and moons bespeak a different kind of life? Simply changing our definition of how life is expressed suddenly and profoundly changes Earth's status in the solar system.

The very act of exploration widens consciousness and compels us to reexamine our assumptions. Early explorers crossed vast distances of land and even vaster distances of sea to discover places, people, and ways of life that sparked in them both surprise and recognition. This same spirit drives us to look outside of our own planet. We find all the outer gas giants festooned with rings, yet the reason for these rings and their formations is different for each planet. Pluto and its moon Charon have recently come into sharper focus and we are beginning to recognize that these two bodies may be a binary planetary system. Looking further out, we catch glimpses of a vast belt of rubble that appears to be the afterbirth of our solar system. Beyond that, we search for the very limits of the heliosphere, that area of space where the sun's energy and its influence shape the way our solar system behaves, and beyond which "outer space" truly begins.

Four spacecraft, finished with their planetary exploration, are slowly moving toward this boundary between our planetary system and the rest of the Milky Way Galaxy. Although they are inert machines, they carry with them a spark of human consciousness. They connect us with the adventure of seeking the limits of the solar system, and ultimately, of humanity itself.

is from the perspective of
e blue planet that we view
of wonder of the night ski
dreds of thousands of point
rkling lights twinkle overh
the night sky. Some are fa
y stars, some are far away
es that are composed of mil
millions of stars, and som
sun's reflection off the c

Index

Photo Credits

Page 8, Jack Harvey / National Optical Astronomy Observatories and Michael Goodman. Page 11, (text box) European Space Agency and NASA. Page 22, 27, Tomo Narashima. Page 46, (text box) Lisa Burnett. Page 52, Don Dixon. Page 70, NASA / Jet Propulsion Laboratory and Alfred T. Kamajian. Page 99, Don Dixon. Page 137, Tomo Narashima. Page 143 Jet Propulsion Laboratory.